BATS AND LADDERS

BATS AND LADDERS

Stories from the Real World of Bat Conservation

GEORGE BEMMENT

PELAGIC PUBLISHING

First published in 2025 by
Pelagic Publishing
20–22 Wenlock Road
London N1 7GU, UK

www.pelagicpublishing.com

Bats and Ladders: Stories from the Real World of Bat Conservation

Copyright © 2025 George Bemment

The right of the above to be identified as the author of this work has been asserted by her in accordance with the UK Copyright, Design and Patents Act 1988.

All rights reserved. Apart from short excerpts for use in research or for reviews, no part of this document may be printed or reproduced, stored in a retrieval system, or transmitted in any form or by any means, electronic, mechanical, photocopying, recording, now known or hereafter invented or otherwise without prior permission from the publisher.

https://doi.org/10.53061/RFJV9716

A CIP record for this book is available from the British Library

ISBN 978-1-78427-580-8 Hbk
ISBN 978-1-78427-581-5 Pbk
ISBN 978-1-78427-582-2 ePub
ISBN 978-1-78427-583-9 PDF

EU Authorised Representative: Easy Access System Europe – Mustamäe tee 50, 10621 Tallinn, Estonia, gpsr.requests@easproject.com

Cover photograph: Common Pipistrelle *Pipistrellus pipistrellus* flying out of its daytime roost in barn roof, Sussex, UK. © Stephen Dalton/naturepl.com

Greater horseshoe bat (page vi) courtesy of Guy Troughton
All other bat portraits by Thomas Bemment

Typeset in Minion Pro by S4Carlisle Publishing Services, Chennai, India

For Neil, Thomas and Matthew

Contents

	Acknowledgements	viii
	Introduction	1
1	The Lord up a Ladder (Part One)	5
2	The Otter in the Wheelbarrow, and Welcome to the World of Bat Work	7
3	Is a Bat a Rodent? Will It Eat Banana? and Other Questions	16
4	Meet the Owners	24
5	Of Carpets and Exploding Nappies	35
6	The Lord up a Ladder (Part Two)	44
7	The Joy of Ladders	46
8	Oh Crap!	56
9	Can I Help You, Officer?	68
10	Batting from a Boat	74
11	The Jacket on the Back of the Door	78
12	Call the Press!	82
13	Teamwork	89
14	I'll Just Have a Look in the Attic	101
	Notes	111
	References	113
	Index	114

Acknowledgements

My thanks to Ruth Swarbrick for editing these stories and for bringing the book up to scratch. My thanks to Peter Chapman, John Haddow and the many bat group members who shared their time and knowledge. Also Henry Andrews, Amanda Barton, Sylvia Bevis, Ian Chadwick, Paul Chanin, Andy Charles, Tim Crawshaw, David Fee, Luke Gibbons, Chrissy Mason, James Mason, Steve Markham, Colin Morris and the late Chris and Helen Shaw. Same to Nigel and Chris Smallbones and David, Helen and Colin Wills. I must acknowledge the contribution of staff and officers of the Bat Conservation Trust, David Bullock and National Trust staff, Local Planning Authority ecologists, Andrew McDouall and other officers from the NCC/EN/NE, architects (a lot of them) and yes, even quite a few of the owners and builders who did their best to get things right. A special thanks to my longstanding and loyal surveyors Anne, Ruth and Sean and to my other helpers. And lastly to Pat Morris, who introduced me to my first British bats as an undergraduate in 1985.

Introduction

There are a huge number of bird books available these days. I know this because I have a lot of them on my shelves and jolly good they are too. But there really aren't that many on bats. Bats are nocturnal in habit, harder to see, harder to study and, frankly, it's taken a long time for the animals to reach mainstream public awareness. Perhaps it's been a case of out of sight and out of mind. They are not everyone's cup of tea, either. You learn this when you have spent over 35 years surveying people's homes for bats.

So this is a bat book. Mostly it's a book about bat work: the good, the bad and the absurd. There's a lot more to working with wildlife than you might realize. Yes, the bats feature, obviously, and the important aspects of ecology and conservation are here as well – these are the founding principles of what we do and why we do it, after all – but mostly it's about the work. It's about standing on the roof of a grand castle on a beautiful dimpsy summer evening watching a spectacular display of feeding activity by noctule bats; it is about chasing a colony of lesser horseshoe bats on their bizarre and crazy flight route through all the rooms and corridors and halls of a fifteenth-century abbey. It's also about misinterpreted late-night encounters with the police, seriously dodgy ladders and a career spent looking at the stuff that comes out of the animals' bottoms. An over-familiarity with all things poo-related, you might say.

I joined the world of bat work in June 1988 with a full-time job on a conservation project. I was trained to identify bats and their roosts and went on to become a trainer for other bat workers, which I continued until the early 2020s. Some of the 300 or so names and faces of trainees remain familiar while others have blurred. Bats have taken me all over Britain and Ireland, to the Channel Islands, Isles of Scilly and Isle of Man, and to Europe. The occupation has come with various titles: bat project officer, voluntary roost visitor, bat worker, conservationist and ecological consultant. Bat

Woman has been a common nickname and someone once called me a batologist, which I thought was rather funny – and I doubt many people know that it actually applies to someone who studies brambles! Technically, I could be called a chiropterist (yes, that is a real word, as in mammalogist for someone who studies mammals or ornithologist etc.) after the scientific group name for bats, but it's a horribly clunky term and painfully similar to chiropodist, which is more familiar to most people. There's way too much room for confusion. Imagine the look on the home-owner's face when you introduce yourself at the doorstep and they wonder why you have come to look at their feet.

This is categorically not a book about bat identification. For most people, even the dedicated bat workers and professionals, identification is problematic and some species look frightfully similar. You can view it as an introduction, though. We have 18 different species. With training, the experts learn to look at facial features and ears, fur colour, the length of key wing bones and the character of the tail membrane. Even their teeth. When the bats are 'on the wing' they can appear slightly different in size and flight behaviour, but more importantly the ultrasonic calls that they make, which are similar in principle to the use of sonar, also vary in frequency and other characteristics and so can be identified from what we call bat detectors.

I like to mentally clump the species into tangible groups. There are the two horseshoe bats – greater and lesser – characterized by the horseshoe-shaped flap of skin that surrounds the nostrils on their face. They are the 'dangly' bats, which hang like pears and plums from their roost perches. Then there are all the others, which include the big bats – the serotine and noctule – and the almost-big bat, the Leisler's. We have two long-eareds, the grey and brown, and a scruffy, pug-faced bat called the barbastelle. There are the pipistrelles, three species to date (common, soprano and Nathusius'), and then there is the *Myotis* family, a challenging bunch for sure. They range in size from the very large greater mouse-eared bat, which you are really never likely to see, and the very diddy little Alcathoe bat. They pretty much all get a mention within the book.

Bat roosts have been mentioned so at this point it's probably best to clarify what that actually means as well. I've been glibly using the 'roost' word for decades, both as a noun and a verb, but for many people it's a hazy thing. What is a bat roost exactly and what does a bat do when it goes to roost? In simple terms we're talking about a feature that the bats enter for shelter and protection and for purposes such as giving birth (a nursery or maternity roost) or for hibernation. In Britain, roosts include a huge range of features and involve built structures of virtually every age, size and character, trees of all sorts, underground sites (natural and artificial) and rocky features above ground, large and small.

A key point is that bats make no adaptation of any sort to any of these features. They don't gnaw to gain entry – they don't have the right sort of teeth to do so – and they don't nest-build. They don't make any structural changes; they merely crawl or fly into the space and settle down and make themselves comfortable. Common pipistrelles are tremendously catholic in their choice of roosts and love buildings of almost any sort, readily use bat boxes and even turn up in rock faces, but you'll hardly ever find them underground. Some species are tree-roosting specialists while others cheerfully mix and match. Our greater and lesser horseshoes bats are the only ones that need flight openings for access.

Bats are typically pretty mobile too, often moving between roosts. In our climate there are seasonal patterns, and different sites and features serve different purposes. A roof space in the summer may be stonkingly warm and great for breeding females and the growth of the babies but less favourable at other times, whereas an underground site which is cool and stable in conditions can be spot on for hibernating animals that want to conserve energy during the winter. Some roost features are inherently ephemeral: think of the gap that appears behind a sheet of old tree bark before it falls off or the gaps that appear in wall cladding and soffits from weathering. Some roosts last for decades, even centuries: think of geological caves, historic buildings and veteran trees.

Overall, there are many roosts that might support casual, intermittent and low levels of bat use each year, but there are a much smaller number that provide all the right conditions of access, space and temperature in which breeding colonies can thrive or in which there may be all-year usage. These roosts, either because of the complexity and range of conditions that they provide and/or because of the stability and longevity of their existence, may be few and far between, but can be extremely valuable to local and even regional bat populations.

Another point worth making is about the way bats reproduce. They do it slowly. They are not at it fast and furious in the manner of small rodents: born precocious, giving birth to large litters from an early age and then keeling over after a short lifespan. Bats take a longer-term strategy. A female typically only has one pup once a year in the summer. This will be in a nursery roost shared with other females – and, depending on the species, some males as well. Good, safe breeding roosts are inherently important. Clearly if one is working in the field of bat conservation it helps to know what species are involved and what sort of roosts you're dealing with.

Most of my work over the years has involved buildings. When it comes to homes we also get to see the most astonishing and occasionally jaw-dropping variety of ways in which people live. We've seen it all, from the stately houses of grandeur to the derelict, sad and filthy piles of squalor. We've seen what's in the attics, not just the grotty, dirty hanging cobwebs but all the stuff that's been put up there over years and years. My goodness, the things that we keep in our attics… the stuff of our youth, of cherished memories and nostalgia; long-forgotten and abandoned objects now useless; empty worthless boxes and absolute rubbish that we have been too idle and lazy to dispose of wisely; bags and bags bulging with clothes, chewed and shredded by the rodents or spoilt with damp and condensation; just plain weird stuff such as bottles of ketchup, a rocking horse… and a stuffed mole.

Like so many occupations, the work has come with ups and downs and highs and lows, but overall I've found bats and their roosting ecology to be fascinating and I've enjoyed the learning. Moreover, I've had great joy from the people that I've met along the way.

It has all been an interesting journey. Welcome to the world of bat work.

CHAPTER 1

The Lord up a Ladder (Part One)

It's always nice to land a job at a place that has its own tearoom. This particular estate sat on the western edge of Dartmoor and had been owned and managed by the same titled family for decades. The garden was open to the public and the buildings currently served a range of functions including ticket entrance and shop, administration, stores and tearoom.

The instructions were to go to the first-floor office where my client – the lord – would be expecting me and where I'd need to start with the roof survey. The airy, open-plan room was occupied by half-a-dozen staff. They appeared moderately but perhaps not intensely occupied. His lordship was delayed, I was informed, so it just seemed best to crack on. The building was long and narrow and the office had been created by raising the ceiling level with sloping sides. This would mean that the original roof space was now much smaller. With my ladder up I put my head through the hatch, and yes indeed, the roof was only about a metre in height. The hatch was at one end: I would need to crawl to the far end where the timbers met to form a hip junction, the sort of cosy, angled nooks and crannies where bats love to roost. This was going to be a hands and knees job.

I started crawling. In such situations the first thing you notice is the complete absence of insulation material and that you can hear every word of conversation in the room below. But progress was fine in the usual dusty sort of way and as I approached the far end of the roof a thin scatter of small bat droppings came into view. Classic pipistrelle signs by the look of things. These bats can roost lower down the timbers from the hip junction, so I leant down on one elbow to twist my head to look back up.

'Fuck!' I cried. There was a huge hornet on its nest 10 inches from my face.

The voices in the office below went quiet.

'Back up! Back up!' my brain screamed, and I went into reverse. It's an inelegant manoeuvre at the best of times, doubly so when necessity requires haste. Nearer the hatch I paused and listened for the angry droning buzz of the hornet. Nothing. It had decided not to attack.

'Ah, hello there, George!' came a voice from behind. 'It's John. How are you doing?' His lordship was up the ladder and was attempting to have a conversation with my backside.

I craned awkwardly to peer behind me. 'Er, all good, thanks.'

There was a pause.

'Anything interesting?' he carried on.

'Just a few pipistrelle droppings,' I answered, slightly strained.

'Is that good?'

Was he really going to continue our little chat in this position?

'Yes, very straightforward,' I managed.

'Splendid. Splendid.' He smiled broadly and dropped from sight.

Following him down the ladder, I tried to muster as much dignity and sense of professionalism as possible.

CHAPTER 2

The Otter in the Wheelbarrow, and Welcome to the World of Bat Work

Not many people set out to have a career in bat work. I'd been brought up on a diet of Gerald Durrell and Gavin Maxwell books and I wanted to be a zoo keeper. For years I pestered the local place – Marwell Zoo, near Winchester – for a job, until they caved in just to keep me quiet. Then I went to the Otter Trust in Suffolk, and finally to a three-year stint at London Zoo. Mostly there was a lot of mucking-out but, as they say, stuff happens. Or at least it did with me. Probably rather foolishly I can claim to have been bitten by the only African spot-necked otter in captivity in the country in 1980 and to have been thoroughly spat at by an angry guanaco. On another occasional a wallaroo escaped and ended up on the bank of the next-door enclosure overlooking the flamingo pond.[1] A squad of keepers slowly encircled it when it made a mad dash for it. Next thing I knew I'd rugby-tackled it to the ground. It was difficult to say who was more surprised, the animal, me or my colleagues. It's not a technique that is mentioned in any of the zoo keeper manuals but the animal was fine by the way.

And once I milked an aardvark, which, it has to be said, is not the sort of thing that crops up on your normal everyday to-do list. A pair of aardvarks was housed in the 'Moonlight World' at London Zoo, not far from the famous Lubetkin penguin pool, which is also famously unsuitable for penguins. One day, unexpectedly, the female gave birth. We certainly noticed but it didn't appear as if she had. Baby, what baby? And she ignored it completely, which meant it would have to be hand-reared. There's a problem, said the zoo vets: we don't have a formula for aardvark milk. Gorilla milk, yes. Chimpanzee milk, yes. Aardvark milk, no.

Cue the keeper who has to get a sample from the mum. So there I was sitting in the den rubbing her tummy, trying to find her nipples. They really aren't very dynamic animals, not the bouncy Tigger-type characters you might assume, and she certainly didn't seem to mind this invasion of her personal space. Even so, it took rather a lot of stroking and nipple squeezing and dodging the odd leg-kick now and then. And that was it: a small vial of aardvark milk which would be enough for the vets to analyse. Sadly, it wasn't enough to keep the baby alive but that's the way of things sometimes when you work with animals.

Lugging an otter about in a box in a wheelbarrow was just another one of those things that cropped up. It had been well documented that our otters had suffered a catastrophic population decline, which started abruptly in the late 1950s following the introduction of the new generation of organochlorine insecticides, such as dieldrin and aldrin.[2] These were used for seed dressings, sheep dips and moth proofing, and proved to be highly effective and persistent poisons. Run-off from fields into the waterways accumulated in invertebrates and then fish and then the predators of the fish. For otters this had a dramatic population effect and by the late 1970s they had disappeared from vast areas of England and south-eastern parts of Scotland. The chemicals were eventually banned and the otter given greater conservation protection under the Wildlife and Countryside Act 1981, but a reintroduction programme was initiated to bolster remaining isolated populations.

The Otter Trust in Suffolk held a collection of animals established by Philip Wayre. At that time he was the only person who had successfully bred European otters in captivity. He also kept several other species including the one and only above-mentioned African spot-necked otter. Scientists from the Nature Conservancy Council (NCC), the government's nature organization at that time),[3] wanted to radio-track some of the captive-bred animals to see how they would behave in the wild after release, but they first needed to work out how to attach a transmitter to an otter. Their idea was to try a soft leather collar, just like a dog collar, to which the transmitter would be fixed, and several nice, quietly spoken fellows, including one called Dr Tony Mitchell-Jones, arrived at the Trust one morning to try things out.

It was my job to bring an otter to the old stables where it would be anaesthetized and fitted with the trial collar. The otters lived in a row of large grassy pens with a free-flowing water channel and one or two large wooden boxes where they could sleep. Each box had a lid that you could lift open and a side entrance that you could shut. The hi-tech operation involved the wheelbarrow. With the chosen otter shut in its box, it was simply a matter of heaving it onto the wheelbarrow and taking it back along the rear path of the enclosures to the waiting team.

Once the otter had been sedated the collar was cut to size and placed around its neck, and when it had woken up I returned it to its pen. In less than 20 minutes the band of leather had been shed and was floating in the river. The team members seemed to take this quite calmly, although I strongly suspected that some of them might have been thinking 'bollocks', it having slipped their minds that an otter has a streamlined neck. Plan B, the design they eventually used, was a harness that went around the front legs.

In 1983, after five years of zoo work, I decided it was time to go to university and study zoology. (I admit this is not the most logical order in which to do things.) The course was particularly strong on mammals and British mammals in particular, thanks largely to my tutor Dr Pat Morris, and as students many of us were regularly brought to conferences and symposia at the Zoological Society of London. I was at one such conference mooching about during the coffee break when I noticed someone I recognized.

'Hello,' I said, 'I don't think you'll remember me, but we met at The Otter Trust.'

The man looked slightly blank.

'I was the one who brought you the otter in the wheelbarrow.'

Tony M-J, as he was often called, became a regular face at other conferences and meetings in the following years. Tony's name would become one of the most familiar and well-referenced names in bat work. Tatty and well-thumbed copies of the *Bat Workers' Manual, 3rd Edition* and *Bat Mitigation Guidelines* are still on my office bookshelf, and the latter is arguably one of the most frequently quoted bat worker references of all.[4] Only in 2023, 19 years later, were the mitigation guidelines updated by P.F. Reason and S. Wray.

On one occasion we shared a rather smelly cave in Portugal. We were part of the British contingent at a mammal conference in Lisbon and had joined a group of local experts on a field trip in search of Schrieber's bent-winged bat. The cave was quite small and relatively user-friendly in terms of access, but definitely smelly, and a part of me wished I'd taken up the option to go birdwatching. The Schrieber's bats were a fascinating sight, however – glossy little animals with heads of satin velvet fur. Sadly, I haven't been back so I haven't seen the species since.

To understand exactly *why* someone should spend 36 years looking inside roof spaces and doing a whole load of other weird stuff, you perhaps need to have a bit more background information, so here it is.

Historically, bats had received relatively little attention. This is surprising when you realize that they make up such a significant proportion of the mammal fauna in the Britain Isles – 18 species at the last count, but see Chapter 3 – but as we have already said they are harder to see and harder to study. From a cultural point of view they also came tainted with some seriously unfavourable PR and were viewed with considerable prejudice, even before the likes of Bram Stoker and the Hammer Horror films swayed people's imaginations. The growing wave of environmental and conservation awareness came in the 1960s. Not only had otters and birds of prey been disappearing but it was also becoming clear that bats were getting hammered as well. Across Europe huge colonies were showing drastic declines in numbers or were disappearing from roosts altogether. All the usual culprits were implicated, including post-war agricultural changes, loss of natural habitat and the use of pesticides, but bats were also subject to direct persecution. The pesticide poisons not only killed the insects that bats ate but also, as an extra bonus, could be cheerfully used to kill the bats en masse while they were roosting in barns and churches and other buildings.

The real driving force for bat conservation in the UK came in the form of the Wildlife and Countryside Act 1981. It has been amended various times and hence is generally and rather unimaginatively referred to as the 'WCA 1981 (as amended)'. The gist of the Act was simple: it defined a whole load of things that were 'offences' such as killing or injuring bats and destroying or obstructing access to a bat roosts. Uniquely, the Act also carried some very specific wording, a proviso that in plain language said: 'if you want to do something that might result in any of the above offences, you first need to contact the NCC and get advice'.

But how did someone in the NCC office know what advice to give? In fact, it was the local bat workers who would make the voluntary visits, work out what the situation was and report back to the NCC so that the formal advice could be issued. The bat workers – the voluntary roost visitors – were trained to be able to tell bat droppings from mouse droppings, to work out what sorts of species were involved and where the bats were getting in and out. Critically, they were also trained in what to say to home owners, how to engage and... when to know to run away. The bat workers were licensed by NCC to be able to make all of these visits. The vast majority belonged to local or county bat groups, and these groups grew in number from a mere 23 in 1984 to 90 in 1992.[5] It's levelled off to around 83 in recent years, including the three from the Channel Islands.

I was late to join the party. It was June 1988 when I was given the job of assistant bat conservation officer for the Bat Project. This was the brainchild of John Burton, the chief executive of the Fauna and Flora Preservation Society (FFPS),[6] and was headed by Tony Hutson, a hugely knowledgeable man, short in stature but fondly remembered for his soft smoker's chuckle. The project was funded by FFPS, NCC and the Vincent Wildlife Trust (VWT). It was set up with various objectives, which included promoting public and professional awareness of bats and bat legislation, particularly within the various building industries that could impact on bats (surveyors, roofers and timber treatment companies, for example), and to provide a supporting and secretarial role for the bat groups within their umbrella organization, which was called Bat Groups of Britain.

My role included liaising with these groups through mail postings and what you might call 'just mucking-in'. It was an ideal position from which to learn on the job and to see the energy and enthusiasm of the group members at first hand. Many weren't content just to do the roost visits; they were busy doing a great deal more than that. And so it was that for a couple of years I found myself with the bods from Avon and Somerset bat groups on an information and display stand at the Bath and West County Show.

In Thetford Forest I joined the Suffolk Group for a bat box inspection and met my first Leisler's bats. Later, there were more of these chunky, scruffy and noisy bats to be seen feeding around the streetlights of Sheffield and Belfast.

In the winter months there was quite a bit of poking around in holes in the ground in search of hibernating bats. One bitterly cold, grey weekend a large team met to carry out a two-day survey of sites in Clwyd, north Wales. Divided into smaller groups, we searched the horizontal adits of old flint mines and scrabbled down caves in the snow-dusted hills. On Saturday night we converged upon a cottage belonging to one of the team where we were generously fed a curry supper (vegetarian or roadkill, take your pick) and offered homemade wines; then we dossed in our sleeping bags wherever we could find space. The gooseberry wine was wicked.

There were more hibernation sites to inspect in Norfolk, this time with John Goldsmith, a man with an irascible reputation blended with a great deal of joviality. Bats are really rather vulnerable during hibernation and disturbance is a major problem, so fixing metal grilles to the entrances to keep people out and the bats safe had become a big thing. Vandalism was common, though, so if you were going to go to all the trouble and expense of installing a solid grille, it helped to have some tricks up your sleeve to keep the vandals out. John Goldsmith knew a thing or two about bat grilles, and his designs deviously incorporated dummy gates within the metalwork that would appear to be weak spots. At one site he positively chuckled as he showed me where vandals had spent a great deal of time and effort attacking the metalwork of the grille without the slightest bit of success.

Bat group members came from all walks of life. Staff from the regional NCC offices (and later English Nature, etc.) were often involved and active (at least while such offices existed, before geographical restructuring and the disappearance of staff into the disembodied ether of centralized headquarters). There were other members who had full-time jobs in wildlife and conservation such as those from county wildlife trusts, city museums and local authorities in addition to volunteering in their spare time, but I can think of a great many more, including teachers, an airline pilot, a roofer, retired butcher, solicitor, phone engineer and a woman who ran riding stables. There was also an artist from the Angus Bat Group, and I have a pencil drawing of his, a graceful redshank in flight, which hangs framed on the wall of my office to this day.

Conservation politics and funding can be bumpy, however, and I had joined during a bumpy period. Money for the Bat Project dropped away and in 1990 a new organization, the Bat Conservation Trust (BCT), was established. It took on the coordination and support of the local groups and quickly grew to become a major organization in the field of conservation.

I continued to work for the VWT, and between 1990 and the end of 1994 we ran the VWT Bat Project from office space at Paignton Zoo in Devon.

The face of bat work began to change in the late 1990s and early 2000s once new legislation in the form of the Habitats Regulations kicked in and local planning authorities got in on the act. These gave extra protection to a range of species, not just bats, and relatively suddenly there was the need for a whole new field of expertise: the ecological consultant was born. In a nutshell, bat volunteers deal with domestic enquiries and the consultants deal with larger projects and those that fall within the planning system. The approvals for domestic works are covered by the advisory system, and other works where bats and their roosts are going to be affected are covered under licences issued by various bodies such as Natural England (NE) or Natural Resources Wales.

It was logical for many of the bat group members to make the move to consultancy. They had the knowledge and experience and were in the right place at the right time. Today a lot of bat workers still wear both hats: volunteer and professional consultant.

From 1994 I carried on doing the voluntary roost visits, the training of new bat workers and running field courses, alongside the growing amount of consultancy work that was evolving. There was other mammal stuff to do as well with badgers and dormice, and mammal courses to teach at Exeter University.

Vincent – the late Honourable Vincent Weir – deserves a special mention. He was a tall, imposing man with a straight smile and a modest demeanour, and most people that were involved with him held him in particular affection. He was a generous philanthropist and was responsible, either privately or in partnership, in funding an extraordinary number of conservation projects across Britain and Ireland, many of landmark significance, as well as supporting a generation of young scientists in their academic studies.[7]

He could also drive people nuts, particularly in later life, when he would often fret about the details of implementation and vacillate about decisions. As an employer he was unswervingly courteous and polite, and I found him to be a kind and considerate man.

He was almost certainly not what we might consider to be 'woke' these days. Born in 1935 and with a certain upbringing, wokeness would have completely befuddled him. Even just my name caused him some consternation. Bemment was my married name, which I was very happy with, but George was a nickname I had had since my early teens and had stuck to like glue because I so hated my christened name. Most VWT staff were issued with business cards. Vincent's first suggestion for my card was 'Mrs Neil Bemment'. I quietly pointed out that this was actually my husband's name, not mine, and we really didn't do this sort of thing any more. He didn't like 'Mrs Bemment' and nor was he happy with just plain 'George Bemment'.

'They'll think you're a man,' he said.

'Surely not if I'm *handing* the card to them in person, Vincent?'

In the end it was agreed that the card could read 'Mrs G. Bemment'. And then, ruminating on my christened name, he beamed and chortled and said, 'Yes, you're probably right. It reminds me of a woman with very large breasts who used to wear a lot of tweed … she was always wearing tweed.'

Thank you, Vincent. I rest my case.

CHAPTER 3

Is a Bat a Rodent? Will it Eat Banana? and Other Questions

Talking to people about bats and answering questions has always been part of the job. Discussions, dialogue and explanations help a great deal to reassure where there are misconceptions and unfamiliarity, and some questions have cropped up more than others.

A common theme has been one of identity, that is, classification. What exactly *is* a bat? 'Is it a rodent?' I was frequently asked, and I guess it seems logical to think that if bats use roof spaces like mice and rats, then they too could be rodents. But no, they're not. Bats are the only order (taxonomic group) of truly flying mammals where the arm, hand and finger bones have evolved to support a wing membrane for proper powered flight. The order name Chiroptera means 'hand-wing'. Do not be fooled by thoughts of flying squirrels or the colugo, also known as the flying lemur. These belong to different mammal groups and aerodynamically they are merely gliders. Of all extant mammal groups, the rodents (Rodentia) represent the largest order, with approximately 41% of all mammal species, and bats, Chiroptera, are the second largest order representing approximately 22% of all mammal species. They span the world from the tropics to temperate latitudes and show astonishing diversity in form. The primates by comparison, which include us (*Homo sapiens*) comprise only 7.8% of mammal species.

And what do bats eat? In temperate Europe all the species are insectivorous, and since all the bats of Britain and Ireland are

European – we don't have any species that aren't found somewhere else in Europe – all of our bats are insectivorous. You need to live in the warmer tropics and sub-tropics if you want to live off a diet of fruits and nectars and that sort of thing. In fact, the variety of diets in these regions is extraordinary.

Our bats definitely don't eat fruit, but on three occasions I've visited people who, having rescued a bat and taken it into care, have then kindly offered sliced banana as sustenance. If you've only ever seen fruit bats in the zoo or in a David Attenborough documentary it's an understandable mistake to make. These days standard and sensible advice is readily available online for bats in need of rescue, and mostly it'll be up to the bat group volunteers to do any feeding that's needed.[1] It won't involve bananas.

What about the number of bat species? How many do we have? This has been a regular question, and the short answer as we've already heard is currently 18, making it the largest mammal group in the British Isles, actually outnumbering the rodents. It's a dynamic thing, though, and the number has changed over the years – so there's also a longer answer, which is what you'll get here. And we need to put things into context. Living on an island means there are fewer species than mainland Europe. Ireland has fewer than mainland Britain, and Scotland has fewer than England since it's a tough way of life for bats in the north. We also need to distinguish between the bats that are resident here and the vagrants that have overshot their normal region of origin. And to add to the mix there's the sad story of the greater mouse-eared bat.

I have a natty little booklet dating from 1986 entitled *Which Bat Is It? A guide to bat identification in Great Britain and Ireland*.[2] It lists 15 species, plus eight vagrants, possible vagrants and North American vagrants. Within a relatively short period after publication the number of bat species had officially dropped to 14, gone back up to 15, then to 16, and now stands at 18. Trying to keep up can be fun, and it may even have changed again by the time that *this* book gets published.

There are a variety of reasons why the numbers change. Take, for example, the fortunes of the greater mouse-eared bat, which have waxed and waned. Small groups of animals were known from a site in Dorset in the mid-1950s and another in West Sussex from the late 1960s, so they probably did breed here, at least for a while. By the mid-1980s the species was represented by only a single, solitary male: Billy No-Mates if ever there was one. In 1992, when this animal could no longer be found, it was with much solemnity that the greater mouse-eared bat was officially declared extinct, effectively the first British mammal to become extinct since the wolf (although no one knows for sure exactly when or where that was). Understandably, it made quite an impact in conservation circles. Then ten years later another individual popped up, and some years later another two. Oops. So the species was back on the list again. However, since these individuals are thought most likely to have crossed the Channel to get here and there's no evidence of breeding, it's considered to be present but not resident. A bit like Spanish expats, perhaps.

In summary, we currently have 17 breeding species... and the greater mouse-eared bat.

Advances in science also explain the increase in numbers. Yes, we've got better at looking but we've also got better equipment with

Soprano pipistrelle

which to do so, including DNA techniques. In the 1986 book, only the common pipistrelle is listed as a resident pipistrelle species, but certainly by the time I started work two years later it was widely accepted that this bat came in two forms: a dark-faced (bandit-faced) pipistrelle that echolocated most strongly at 45 kHz, and the other, a fractionally smaller and paler animal, that echolocated most strongly at 55 kHZ. We called them 45 pips and 55 or blond pips (what we now know as the soprano pipistrelle). It took field studies to confirm that the two types had different roosting and foraging ecology and the DNA science to confirm their genetic identities.

The Alcathoe bat, which was only added to our list in 2010, was also pinned down through genetics. It's a member of the *Myotis* family. It looks ridiculously similar to its relatives, the whiskered and Brandt's bats, and had perhaps been hiding in plain sight. I've never seen one or if I have I haven't known I have. If that makes sense.

For the Nathusius' pipistrelle, it changed from being a vagrant to a resident breeder, not unlike the little egret in the birding world. Researchers realized that instead of occasional occurrences, zillions of hours of sound recordings were beginning to show that the bats were flying in well-defined peaks of activity in the spring and autumn months, principally off the east and southern coasts facing Europe. And then males were recorded in mating song flights in parts of the country, and hey presto, a growing number of nursery roosts were identified.

The calls from Nathusius' bats would occasionally turn up on sound recorders that I was using for Devon sites, especially in the south of the county. On one occasion a fisherman phoned from Brixham harbour to say that a bat had dropped down onto his boat while out at sea. He said he thought it looked 'knackered'. His exact description, I promise. I collected it and sure enough it was a Nathusius' bat. It was fed, watered and later released. Over the years I've found several transient individuals in buildings and it's always been a surprise. The brain goes: 'Oh, we have a pip here … no it's not … oh, a monster pip? Better measure it.' They do indeed look like a pipistrelle but in a noticeably chunkier and shaggier sort of way, particularly in the autumn.

In the summer of 2023 I collected a sample of pipistrelle-like bat droppings from the roof of a small detached bungalow, 6 miles from my home. They appeared just ever so slightly different from the norm. DNA testing confirmed them to be from Nathusius' bats. That was a first for me: the first time I had ever knowingly found Nathusius' bat poo. Changing times, obviously.

While talking about Nathusius' bats I'm reminded of another strongly migratory species, which is the Leisler's bat. It's a resident breeding species with a patchy distribution in England, south-west Scotland and a stronghold in Northern Ireland. It doesn't seem to like the south-west or Wales much and I've never come across a Leisler's roost in Devon, but I have come across two grounded animals. The first was found one morning in April on the playground of a nearby primary school. It was terribly thin and weak and sadly died the same day it was found. Many years later, on almost exactly the same date in April, another bat was found, bizarrely in the lingerie department of a Newton Abbot department store. It was fit and feisty and was delivered to my front door in a cardboard box while I was making tea for my two boys. It was calling loudly, and when I glanced very briefly in the box (multitasking), I took it to be a noctule bat. So we christened it Norbert. After tea and closer attention I realized I was wrong and so we had to rename it Larry. It was, as standard, fed and watered and later released.

The point of the story is not the lingerie department, although heaven only knows how that happened, but the condition of the first bat and the timing of both occurrences. To my mind, both animals had been migrating. The first one didn't make it but plenty do. Individuals of other European species have turned up on our shores. The parti-coloured bat has dropped in, most often during migration periods. Kuhl's pipistrelle, Savi's pipistrelle and the pond bat have been identified as well. When found grounded in counties in the east and south of the country there's a reasonable chance that these bats have made it across under their own steam, even if perhaps they didn't intend to do so. With climate change, who's to say that one or more of these might not follow the example set by the Nathusius' pipistrelle and eventually start breeding here?

Some bats very clearly get travel assistance, such as the young Savi's pipistrelle that was discovered in Scotland in a box of nectarines imported from Italy. And then there's the small number of American bats that have turned up here. The majority of these have almost certainly hitched rides on boats of one form or another, although strong fliers like the Hoary bat could perhaps make it over on their own if wind-blown.

A subject that frequently prompts questions is about predation. What animals hunt our bats? The ladies of the WI have often seemed particularly (morbidly?) interested by this, and it is indeed an interesting question, something you can really get your teeth into if you're so inclined. Most people first think of tawny owls or barn owls since they fly at night and hunt small mammals. In fact, these birds are best designed to hunt terrestrial prey such as rodents and shrews and not something that's on the wing. That doesn't mean to say they won't nab a bat if they get the chance. Very occasionally bat bones certainly do turn up in tawny owl and barn owl pellets, the undigested, regurgitated remains from the bird's meal. And both greater and lesser horseshoe bats are known to have abandoned roosts if a barn owl moves in, which strongly suggests a desire not to be predated.

Flying at night is a jolly good way to avoid predation from the daytime birds of prey, such as the sparrowhawk and peregrine which are specialist aerial hunters, and for the most part it works. But again, some of these birds are adaptable and opportunistic and learn to beat the odds. A sparrowhawk near Bristol once learnt exactly what time the greater horseshoe bats started to emerge from their roost – a time when it was still relatively light – and took to making a perfectly timed fly-by each evening, grabbing one in the process. Given the importance of the roost, the conservationists in charge really didn't think this was very 'PC', and were allowed to trap and move the bird.

Examination of the feeding remains from a pair of peregrines nesting in a church tower in central Exeter revealed that they had caught at least one noctule bat. They had also been catching some very weird bird prey, the sort of birds that were more likely to have been flying at night on passage than during the day, all of which strongly suggested night-time predation against a moonlit sky. Remarkable.

Kestrels are not normally associated with hunting bats, but I've seen one hovering at dusk outside a large stately home, waiting for the pipistrelles and brown long-eared bats to emerge. It did so for several evenings, and although we never saw it catch anything it wasn't for want of trying. A colleague of mine once observed a hobby – a proper aerial specialist – chase and catch a bat with more success.

The bats' way of life means that they avoid predation from the majority of our native mammalian predators. There are exceptions of course. At Nietoperek in western Poland there's a long, long system of deep concrete tunnels dating from the Second World War. Post-war, when these were abandoned, bats from all over Europe began using the tunnels as a hibernation site and conservationists and researchers now regularly count around 35,000 bats or more each winter. With such a mass of small bodies it's not a surprise that stone martens (a close relative of our pine marten) also visit the site and nosh the odd bat. Most of the bats roost out of reach and the martens scavenge for corpses, but scratch marks on the walls and the identification of bat bones in the marten scats also show they're actively hunting.

Much to the dismay of the WI ladies, the reality is that the bats' greatest predator is the domestic cat. Many cat owners have probably guessed as much already. Cats are hotwired to hunt, tend not to be fussy about what they hunt and there are a very, very great many of them. I Googled it and the figure I kept finding was 10.8 million in the UK. I reckon at least 18 of those live in the short road where I live.

The maternity roosts where the greatest numbers of bats gather at one time attract particular attention. A cat may not be able to get into the roost but if it can climb *onto* the roof or get to an eaves position it can swipe the bats as they emerge. Some are very proficient and can injure and kill a significant number of animals before the harm is discovered. In other cases, the cat brings its prey back home and drops it on the kitchen floor, where the owners discover the body is that of a female with a new-born pup attached.

One or both may already be dead, but if not the chances of survival are generally very poor.

In such distressing situations the first measure is to keep the cat under curfew at night until the roost can be found and measures taken to protect it – which sounds perfectly reasonable. Unfortunately not all cat owners think so. One woman, who had reported that her cat had been bringing in dead and dying pipistrelles – mothers and babies – for almost a week did initially agree to curfew her cat for several days. Very sadly, after we'd failed to find the roost in that period she then refused to keep it in any longer, and it went back to hunting and killing some more. We were well into double figures by the time she stopped talking to us.

Bats also get whacked by wind turbines, big and small, just like birds, and finally, yes, they get killed by car strike too. It's inevitable. A good friend and colleague took a call to collect an injured bat one evening. As he was driving to fetch it, a flying bat hit his windscreen in an obvious instant-death sort of way, much to his dismay. The injured bat he'd gone to collect had also died by the time he got there, so it wasn't a good evening overall.

CHAPTER 4

Meet the Owners

It's difficult to say quite how many home owners one might have met during a 36-year career of surveys but it's quite a big number. Which means by the law of averages one's probably met most sorts. I've never knowingly met a mass murderer or hardened criminal but that's not to say that I haven't met them *unknowingly*. Some people certainly come across as being shadier than others, and there have definitely been a few property developers who have given every indication of being out-and-out crooks.

There have been some gob-smackingly rich people, most of whom behave like perfectly decent human beings while the others behave like complete pillocks (that's the most polite word I can think of). I can think of several who possessed such an overwhelming sense of entitlement that they clearly no longer felt normal good manners and abiding by the rules applied. There have been oddballs and eccentrics, and those who unfortunately, for one reason or another, appear to have lost the plot.

I've never met anyone who was buck-naked, but my tall, lanky friend and trainer Peter Chapman did.[1] Peter describes being called to the home of an older couple who were naturists. He was invited to sit around the kitchen table drinking tea and chatting with them while they sat there wearing next to nothing. 'I've never spent so much time admiring an Artex ceiling,' he says. He can't remember a thing about the bats; he just remembers the lack of underwear.

And that's the thing: there are some roost visits and survey jobs you just don't forget, not because of the bats but because you remember the people – the frailty, the humour, the unpleasantness or, conversely, the warmth. You remember the emotional clout that came with the situation, be it deeply depressing and infuriating or enriching and uplifting.

Bereavement, old age or other sources of anxiety could be the subtext for many of the calls we received as voluntary bat workers. One woman from August 1991 stands out. She lived in a 1920s brick-built, pebble-dashed house on the outskirts of Ashburton, on the southern edge of Dartmoor, and she had bats flying into the bedrooms at night. Her husband had only very recently died and now she was raging at the bats. She said they haunted her. They'd been in the roof for many years and had never been a problem while he'd been alive, because he'd always been the one who dealt with them and who let them out. But now here they were, reminding her that he'd gone, and her grief and her distress were obvious. She wanted them to be removed immediately. 'Just get them out of my house,' she sobbed.

Unfortunately, there were at least 15 brown long-eared bats in the roof on the day of my first visit, including some youngsters, so even with the best will in the world we were unlikely to get approval to exclude them at that time of year. She'd asked for help and we would do so, but we could only do so in a certain way. I came back with Peter the following evening to see where the bats emerged from the roof – almost all from gaps at the inner edges of the deep overhanging soffits – and we discussed options with her. In the end we came up with the idea of having three meshed frames fixed to the bedroom windows to keep the bats out – and the VWT Bat Project paid for these.

We hoped that this might suffice, and indeed by mid-October she said she was happy with the mesh. But no: the following January we were instructed to do a formal exclusion. So we returned, and she hid downstairs while we searched the roof, captured the two bats that we found inside and moved them to a small adjacent barn, before finally stuffing cloth rags into the soffit holes.

A Mr and Mrs Mills lived in a 1960s house in the village of Marldon, and they too had had a bat – a common pipistrelle – fly through an open window, in this case through the dormer window, where it then did a perfect RAF fly-by down the stairs and into the living room. They had seen tell-tale droppings on the window sill of the dormer for several summers, but had never quite worked out what they were from until the bat had overshot its normal landing mark.

I went over to extricate the animal from the living room curtains and, as we did so often, to chat. But it was Mr Mills who did the talking. He believed the house was suffering from subsidence, that the roof had shifted and that it might blow off in the next gale; his dog had died recently and there were money problems. On the roost report form for the visit, I wrote 'Mr Mills possibly suffering from anxiety problems.' Almost certainly.

Just down the road from Mr and Mrs Mills was an elderly lady who called because she said she could hear the bats in her bungalow. It turned out to be her smoke detector telling her that she needed to replace the batteries, but in the very short amount of time it took me to work this out she'd already put the kettle on for a cup of tea. There was an awful lot of talk about her general state of health and then she said, 'And I have a pacemaker, you know, look, just here.' And she grabbed my hand and held it to her chest. This wasn't bat work, it was more like social care – and it wasn't hard to see the loneliness.

The call-outs from angry, hostile owners were more challenging, and you never quite knew for sure how things might pan out. One such call was from a Mr Ingram. He had phoned the local wildlife trust to say 'someone needs to take these bloody bats away or I'll kill 'em.' It was July, so he almost certainly had a breeding roost.

There were two of us to make the visit, always the best approach for a potentially difficult situation, although we probably didn't look very robust. I was six months pregnant and my companion, Alison, was a lovely, smiley and very petite young woman. Mr Ingram, on the other hand, was a big man and downright belligerent. And goodness, how he liked the sound of his own voice. He jabbed his finger towards the tile cladding at the front gable of the bungalow and the bat droppings on the window sill and the ground below. When he wasn't expressing himself we could hear the bats faintly chattering to one another at the top of the wall.

With the utmost consideration we explained that, no, we didn't need to go into the roof and cause any domestic inconvenience, but we would just put a ladder up to the outside gable and use a net to catch one or two bats as they emerged, 'just to confirm species'. We didn't really need to do that either, since we could tell from the droppings that this was a pipistrelle roost, and our bat detectors, which could 'hear' the bats' ultrasonic calls, would also have confirmed this. But as they say, a bat in the hand has its uses. Actually, no one says that.

The bat net basically just looks like an oversized butterfly net on a long pole, and Alison was very nifty with this. It's not normally a spectator sport, you understand, but other members of what appeared to be an extended family clan began to appear and loiter and watch for a short while before disappearing back indoors to the telly. Mr Ingram's tone became slightly less hostile and there were even some attempts at humour. We showed him an adult female pipistrelle bat in the hand and he seemed interested, even in a vaguely pleasant sort of way. Alison brought down a small, newly flying youngster as well.

The evening stretched on. We'd been there for over an hour taking his barrage of provocation and commentary, and we'd said and done as much as we could to explain and inform. It was time for him to tell us what actions, if any, he wanted to take regarding the roost so that we could report back. Nah, he said, he wouldn't do anything: he'd changed his mind and would be happy to have the little blighters; he didn't want to hurt them, especially not the babies. And then he announced that it was a hobby of his 'to aggravate people' and we'd been good sports, so thanks very much for the visit. We thanked him too for his time, went back to the car and drove off round the corner where, out of sight, we pulled over. What a lovely man we both agreed, what a pleasure…

When it comes to eccentrics, Alan Matthews was the best. He lived in a gorgeous, thatched stone cottage near the Dorset border and he had bats and a grand piano. His front door opened into a spacious hallway with bare stone flooring, and the grand piano sat in pride of place in the arc of the curving staircase. He gave us the full-blown house tour, waving his hands with a flourish to indicate stonework that dated as far back as 1380. He had been cheerfully living with free-range lesser horseshoe bats in the house for the last 30 years. They mostly lived in the thatched roof space, but a part of the central top floor ceiling had collapsed and he had never quite got round to fixing it – so the bats had moved into the hot water tank cupboard, and he had even given them extra space in his wardrobe, moving some of his clothes to another room.

In the evenings they flew up and down the landing and came and went from the house through one or more bedroom windows, which he kept open for them. They seemed to prefer his bedroom window, he said. And when he had guests and he was sitting playing the piano at the foot of the stairs, they'd flutter down and around and investigate before departing into the night.

Twenty-four years later I was asked to survey the house once again. The address was unfamiliar, it had been too long ago, and the name of the owner was new, but I recognized the hallway. It was all that remained. Alan Matthews had long since passed away, the roof had been mended and the piano and the bats were all gone.

Two other people that I remember with fondness were a Miss Georgiana Ellis and her sister Brenda. They were a delightful couple of old ladies who lived in an elegant town house in Tiverton not far from the river. The house had been built in the Queen Anne period,[2] to which a rather incongruous double-height, flat-roofed extension had been added at a slightly quirky angle in the early 1970s. Brenda's bedroom was under the flat roof and for several years this roof had been occupied by a large summer colony of soprano pipistrelles. The sisters had counted out 230 bats the previous year, and they were fiercely protective and proud of 'their' bats.

They didn't mind the squeaking and chattering, and they didn't even mind clearing up the droppings that were liberally and conspicuously showered outside on the window sills and patio below. Recently, however, things had got a little out of hand. They had a 'smelly roost' and poor Brenda had been forced to decamp and sleep in another part of the house. The room stank of ammonia. There was no way one could play it down or pretend 'Oh, it's not too bad.' It really was bad … bad like an old-fashioned French urinal.

Believe it or not, most batty roofs don't actually smell that much. They may be a tad musty but not what you might call properly horrid, and that's either because there are relatively few droppings – because there are relatively few bats – or because the droppings that do accumulate are in a spacious and airy void where they mostly dry out. For Georgiana and Brenda, their bats had been peeing and pooping in a very confined space, and to make matters worse their builder said the flat roof was leaking where it had been built onto the main slate slope. Whatever the reason, the resulting chemistry was truly eye-watering.

And yet they were both wickedly cheerful and friendly about it all. Yes, they would like a solution to the smell problem please, but without any harm to the bats, and would I like scones and jam with my tea? It was agreed that I would go back and do an evening count, which included an invitation for supper. For the sisters, you see, this was all one gloriously big adventure, and when the count that July revealed they had over 400 bats in the roof they were absolutely thrilled.

In the following November, forms having been completed and permission granted, the builders and I took the flat roof off. We hired an industrial hoover, paid for once again by the VWT, and removed the accumulated droppings. There really were quite a lot. A new flat roof was put back on with the access gaps as before and ventilation discs installed within the fascia boards. Bats did return over the next couple of years, although not the colony en masse, much to the sisters' disappointment; but Georgiana kept in touch and I was again invited to supper. I still have the recipe for her avocado starter.

The relationship between a voluntary roost visitor and an owner is very different from the relationship between a bat consultant and an owner. Both turn up in order to assist the owners with their bats, to help resolve problems and to enable them to do what they want to do to their property within the appropriate planning and legal frameworks. The consultant comes as a paid professional, however, and in the vast majority of cases it is the local planning authority that's telling – requiring – the owner to employ said professional. Hence the dynamics may be unfavourable from the outset: we're being employed by someone who might really, really not want to employ us.

The variety of owners is still the same, varying in temperament from totally engaged, upbeat, helpful and thoughtful to downright bolshie and morbidly anti-bat. Or anti-wildlife of any description, come to that. But the potential for conflict is definitely greater when bureaucracy, seasonal timing issues and spiralling building costs are thrown into the mix. As with any profession, consultant ecologists are bound by published professional guidelines and protocols, and certain surveys can only be done at certain times of year, for example – but that can be a difficult pill for owners to swallow when it means delays to their projects.

It's not the consultants' fault that bats and their roosts are protected by law, but some owners I've encountered have maintained an almost permanent state of resentment, occasionally boiling into outright anger. The amount of energy harnessed for such a state can be a tour de force, but it makes for a difficult and unpleasant working relationship. There was one exhausting woman from north Devon whose planning application included demolition of part of the house and replacement with a new extension. The demolition bit included a range of features used by bats, and the law allowed her to demolish those roosts under licence. By the time the licence was obtained and the works had started, she'd abandoned any semblance of entente cordiale.

On our last encounter I'd called by to see how the builders were getting on with the replacement roost detail that had been agreed. She screamed at me like a banshee for the temerity of visiting her – even though it had been clearly described as part of the process – and screamed at me some more to make me leave, which I did hastily. The builder came puffing and panting up the drive behind me, out of breath but desperate to say he'd do what he could. Poor chap, he still had to work for the woman. Mercifully I'd just been sacked.

There was another Mrs Angry, this time near Plymouth, who was also a shocker. She and her husband were spending a fortune to have one house demolished and a new one built in its place. It didn't seem to matter if the subject of communication involved reporting, timescales or completion of the necessary forms, her go-to expression was barked down the telephone and written Trump-style in capitals in her emails: 'MAKE IT HAPPEN, GEORGE! MAKE IT HAPPEN.' It was very wearing. Who wants that yelled at you the whole time?

Exceptionally, there was one relationship with a property and its owners that spanned a 17-year period during which time I visited both as a volunteer and as a consultant. Ken and Patricia Gurnell bought the East Court Hotel in 1998, a Grade I listed building in the heart of the Devon countryside. It was the most wonderful building, a fourteenth-century manor house with Elizabethan and Victorian extensions, and it came with adjoining stables, an orangery and other outbuildings, as well as a walled garden. It had been run as a hotel since the 1970s, but it was big, and almost all the buildings needed attention in one way or another.

The first thing they needed to do was to have the house re-roofed and treated for woodworm, then there was an eighteenth-century French bell and its bell tower that needed restoration, repairs and reinstatement of the original Victorian kitchen, conversion of the redundant laundry rooms and yet more roof repairs to the staff cottage. All of this required juggling not only the demands of Listed Building status but also the conservation needs of the bats that used the place. It was heaving with them. The hotel roof was home to a resident population of lesser horseshoe bats, which included a large summer breeding colony, and there were at least seven other species that had spread themselves around as well.

Ken dealt with most aspects of the building projects and he had an extraordinarily enlightened attitude towards the bats' presence. Scaffolding surrounded the hotel during an early stage of work but it remained open to guests; hence a detailed information display was produced for the entrance hall. Ken specifically requested that it should include illustrations of the bats and the roosting details that we'd collected by that time. It even included a diagram of their convoluted flight route from the central roof chamber, which served as the maternity roost, through openings in brick walls and interconnecting voids, through the hole in the collapsed ceiling of the Victorian kitchen, past the old bread oven and out through the coal cellars.

He'd often sleep over in the small staff flat on the very top floor. It had a kitchenette and a mini-balcony where he could stand and smoke. It also had a somewhat incomplete hatch door to a passageway that led to what would have been the old servants' bedrooms. These were derelict, and there were various gaping holes in the lath and plaster where the lesser horseshoe bats would appear and occasionally make their way into the flat. He wasn't in the slightest bit fazed by this.

He and his gardener were hugely enthusiastic about the droppings that we shovelled up from the maternity chamber every few years. Bags of it. It went as fertilizer into the vegetable beds in the walled garden.

'Leave it by the back door,' he'd say cheerfully. 'Charlie will come and get it.'

Charlie, a lovely white bearded fellow, swore by the stuff.

Several things stand out in relation to the Gurnells' 17-year stewardship of the property, one of which is how well the bats fared during the period. You might have been somewhat apprehensive given the amount of roof works and alterations that took place. But almost every summer between 1999 and 2014 we counted the number of lesser horseshoe bats as they emerged and this went up from a mere 133 to a peak of 270. Admittedly the figures reflect a general, nationwide increase in the population, and a jump of 50 bats one year probably included the arrival of an extra bunch from somewhere else, but the point is the trend was always upwards. The number of greater horseshoe bats using the cellars increased over the years as well.

The other things that stand out were the warmness with which we were greeted whenever we bowled up and Ken's generosity. You'd expect practised hospitality from someone who runs a hotel, but it always seemed to go beyond that. The various building works required numerous site visits, sometimes for meetings with the architects and contractors, and sometimes things just cropped up. Invariably we'd be offered refreshments: coffee and the hotel's own homemade biscuits or petit fours mid-morning, sandwiches nearer lunchtime, tea in a teapot and more petit fours or cakes later in the afternoon. *Foie gras* pate was even proffered on one occasion, which was far too good to resist. Wine was frequently offered, although mostly and reluctantly declined.

Ken's generosity was expressed in other ways too. He and I were sitting at the bar one day, by which time we'd recorded eight bat species from the buildings but hadn't done any surveys of the grounds as such. We got onto the subject of hazel dormice. My hunch was that these would be dotted about as well; it was that sort of place. I explained that I also did dormice surveys if he was interested, and then, just because the conversation had taken us in that direction, I mentioned that there was to be an international dormouse conference in the south-west the next year and colleagues were trying to organize venues for field trips.

Without hesitation, he offered cream teas for 60 delegates. It sounded fantastic, and I asked how much he'd need to charge so it could be added to the conference budget.

'Oh, that won't be necessary,' he replied. 'It can be from me.' The conference was held in Somerset in the end and this offer was never needed, but he had meant it.

Not long after this one of my trainees put a foot through one of the bedroom ceilings in the hotel. From the beginning Ken had given his blessing for occasional training visits with would-be bat workers, National Trust staff and other professionals needing roost experience. Trainees regularly got to help with the summer colony counts, but the hotel and its buildings were perfect for daytime inspections as well. There was easy walk-in access to the cellars where at least some bats were almost always visible, there were the stables where other bats and signs of use could be examined, and there was the narrow passage from the staff flat where we could peer into the maternity chamber. The passage *was* narrow and low, but everyone always got the same briefing.

On this occasion I'd been with three trainees, we'd done a full day and I was back at home when Ken telephoned to say they'd found a hole in the ceiling. It couldn't possibly have been my group, I said, horrified. No one had mentioned anything; there must be a mistake. Except it was the only explanation. I drove straight over and one of the staff took me up to the room. The foot-sized hole had been made just inside the passage from the staff flat where the four of us had been earlier in the day.

Ken was walking back from the orangery when I met him on the lawn. Professionalism went out the window and I burst into tears. I was mortified. This was the only time it had ever happened anywhere, and it wasn't just a hole in any old ceiling but a listed building ceiling – and my favourite, favourite roost of all time. I sobbed that my insurance would cover it, and I repeatedly apologized. He gave me a hug and kindly told me not to worry.

The hotel covered the damage on its own insurance in the end and continued to seek advice regarding other building work, but I had no heart for any more training visits. No one puts a foot through a bit of lath and plaster ceiling without knowing that it's what they've just done. No one admitted to it, though.

In 2015, shortly before getting organized to do the annual colony count, I learnt that the hotel had been sold. A new company had taken over and new ecologists had been appointed. They never did make contact to find out what we had learnt from the last 17 years, and sadder still, I never got a chance to say goodbye.

CHAPTER 5

Of Carpets and Exploding Nappies

Before reaching the loft hatch and the attic you have to cross the threshold and make your way through the 'living space' of the home, of course. So at the very moment the front door or side door or whatever door is being opened to you, you're already making the following assessment: am I going to need to take my shoes off?

It goes something like this. Is the person who comes to the door wearing shoes? Or are they in slippers? Is there a blatantly obvious shoe rack cascading with the family shoes next to the doormat? And, most importantly, what's the state of the carpet? Does it appear lived in and laid back and comfortable in its existence or is it spectacularly pristine and scary?

Like everything else in our homes, what we walk on varies enormously, and for someone who has to walk through the homes of other people on a regular basis the state of the flooring can be a surprisingly stressful element of the job. In a bungalow the route from the front door to the loft hatch may be short and pain-free. In a magnificent Victorian villa in the heart of Dartmoor there may be multiple loft hatches on multiple floors accessed only via polished marble flooring and spotless, tasteful colour co-ordinated carpets and expensive rugs. They all have to be navigated with care and caution, and occasionally a very strong stomach.

At the pristine end of the spectrum was a house on the outskirts of Chudleigh that needed a survey. I should have twigged some clue about its character when I was asked to liaise with the housekeeper for access. It was surrounded by a stark and intimidating brick wall with railings, as if it had been picked up from somewhere like Sunningdale and plonked in the wrong county. You had to punch in a security code to get through the heavy spiked metal gates, and even the huge polished wooden front door looked unfriendly.

Eventually the housekeeper opened this door. Behind her was a pale shiny floor and behind that a wide staircase with a white carpet. Not off-white, not creamy-white. I mean proper white-white. I'd never seen a carpet like it.

'Do you need to look upstairs?' she asked pointedly.

'Yes, please,' I said. 'I'll just get my ladder and things from the car and then take my shoes off.' Somehow I sensed that this would be a sensible statement of intent.

We can and do wear the little plastic slip-on shoe covers, of course, but mostly they're a faff and not necessarily a good idea on ladders or in the roof itself. They're also difficult to recycle, so many of us prefer the shoes off – shoes on – shoes off – and shoes back on again approach. In this case a dust sheet was obviously also going to be obligatory.

The housekeeper led the way upstairs and along the landing to the ceiling hatch, and then signalled her complete lack of interest by disappearing. The carpet was spotless. The pale paintwork was also spotless. Lifting myself into the attic, that too was spotless; there weren't even the usual spiders and dead flies. It was a sterile, hermetically sealed roof. Back on the landing again, I noticed several of the bedroom doors were open. Clearly this was a family home: these were children's bedrooms. But where were the usual tell-tale signs of children? The house was immaculate and clutter-free. I could understand if there'd been a bit of a tidy before my visit, but where were the spillages and marks on the carpets? Where were the smudges and scrapes along the staircase and walls? Children are messy by nature and the livelier and more exuberant they are the messier they are. Were these children banned from moving when eating and drinking? Was the consumption of anything remotely chocolatey restricted to outdoors? Ketchup – was this under lock and key? And ice creams, were they a complete no-no? Did attempting to raise children in a house with a white carpet amount to child cruelty? Almost certainly not but it struck me as very weird indeed.

We tried our best not to make a mess on the carpets during our visits. Owners are usually perfectly capable of doing this themselves. I immediately thought of a very nice retired couple who lived in a gorgeous granite and thatched cottage next to the village church. The entrance hall and living room still had the old stone flooring. Tassel-edge rugs were dotted about here and there and a light aquamarine-coloured carpet led up the staircase and along the landing. For quite some years the couple had known that there were bats in the roof, and they were happy with this. However, in recent months some of these had taken to flying down the chimney in the early evenings, emerging from the fireplace and dangling and twiddling their thumbs, so to speak, underneath one of the large lampshades in the corner of the room.

This behaviour was disconcerting at first, but the couple were more intrigued than troubled and the husband was particularly keen to get into the roof to learn more. The ceiling hatch opened above the landing. The ladder and dust sheet were all brought into play, and he was up and in the roof like a shot. It was largely boarded out, which made it easier to investigate, and he was entranced by the sight of several small, pale, sooty-grey lesser horseshoe bats roosting quietly under the roof slopes. So entranced was he that he also seemed oblivious to the generous and conspicuous spread of droppings underfoot. This was definitely a well-established breeding roost, and there were lots of them. We identified the gap at the eaves where the bats got in and out, and closer inspection of the chimney showed the break in the stone column that was allowing the more adventurous animals to go exploring downstairs.

Satisfied we'd seen what we needed to see, I was back at the hatch first, so dropped down the ladder and put my plastic booties on before stepping off the dust sheet. The wife was waiting on the landing to hear about our findings. Her husband came down behind me and gleefully marched off towards the stairs… leaving the most perfect set of bat poo footprints on the aquamarine. His wife looked at him, then at me, then back at the carpet. She opened her mouth but no words came out. At least not until after I'd left.

It's not just the carpets that we have to worry about; there's worse. Picture an ordinary semi-detached house in an unattractive street in a poorly planned and ugly town. I was on a training visit with Peter Chapman. The issue apparently involved complaints from the neighbours about a bad smell that, it had been alleged, could have been from bats. But just walking up the path towards the front door it was kind of obvious that that wasn't the issue. There was definitely quite a whiff, and it wasn't a batty whiff.

The owner, a short lady of indeterminate age with dyed black hair and a strained expression, opened the door to a narrow front hall from where the tight staircase started upwards on the left, an open doorway led to the kitchen ahead and a second doorway to the living room on the right.

The floor – extending from the base of the stairs and into the kitchen – was covered in litter trays and sheets of newspaper, all splattered in cat poo and brown urine stains. Cats of various colours mooched about at our feet. The place stank. That wasn't all. The staircase was stacked with bulging black bin bags. From the bottom turning point to the small landing at the top, bin bags lined the wall and at the top of the landing there seemed to be a mountain of them. What looked like dark manky cobwebs decorated the wall tops and the ceiling corners.

It was disgusting but we couldn't say anything. We just wanted to do what we had to do and leave as quickly as we could. We squeezed our way up the stairs to get to the hatch – the edges also lined with gunky cobwebs – where we were then faced with the problem of all the bags. There was no space to get our ladder up. Unwisely we opened a bedroom door, only to find the room was also full of bags. We had no choice so heaved those from the landing on top of them. Peter went up into the roof and I stood and peered in from the top of the ladder. It was hanging with thick matted and trailing cobwebs, and skeletons and partly feathered corpses of dead birds littered the ceiling timbers. This wasn't a bat roost, and it wasn't a situation that we could resolve.

Fifteen years later I was with another good friend and colleague, Paul Chanin, when we encountered a similar scenario that was even more unsettling. The survey was for an extensive housing development on the edge of Plymouth; it included a large number of buildings ranging from stand-alone bungalows to whole farmsteads. The work required a two-person team for each survey day, and it was day three with only two more properties to go.

We drove up the muddy and bumpy track to this particular farmhouse. It didn't look anything special. Not an old, traditional stone-built structure and not new, just dirty and worn-looking, square with a cheap timber-framed and glass-roofed porch at the front. There was no lawn: the bare surface of the drive simply extended to the porch and off to the right, where we could hear cattle mooing and coughing in a stock barn.

The porch door was locked and there didn't appear to be a door bell. We peered in through the windows. Plastic supermarket bags stuffed with goodness only knows what filled the space, piled high against the back wall and beneath the window sills. We could just make out a narrow route between the bags that led from the porch door to the inner door of the house. There was an unspoken sense of unease. We went in search of a back door and found a small conservatory where the windows again drew us closer. A cheap metal-legged table and a couple of deck chairs were just visible beneath more plastic bags, while dozens and dozens and dozens of empty pet food tins littered the space. The jagged edges of open lids stuck out alarmingly. There was nowhere to stand or sit in the conservatory; it had become another dumping ground.

We went back round to the front of the house, arriving just as the farmer appeared from the direction of the barn. He was a short, round-faced man who immediately smiled and stuck out his hand in a lively, friendly greeting. He was wearing what had once been bottle-green coloured overalls but were now a dirty autumnal brown: the front was rigid and polished with animal grease. It didn't look as if the overalls had ever seen the inside of a washing machine.

But as I said, he was very friendly. He opened the porch door, all the while chatting with a broad Devon accent and telling us about the farm. He paused just inside the door to slip out of his wellington boots and put on a pair of slippers, and then he led the way past the mass of garbage (as if it wasn't there) and through the inner door. This opened directly into the kitchen. We didn't pause but went straight through to the opposite door, which led to a tiny hallway and the bottom of the staircase.

It was a good thing we didn't pause. The kitchen had a Formica floor typical of the 1970s that once upon a time had a pattern of large black and white squares. The pattern was only visible in the short, well-used route between the two doors. The rest of the floor was covered in a layer of filthy, filthy brown grime. The kitchen surfaces were no better. Every surface was strewn with dirty frying pans, saucepans, crockery and cutlery, and the remnants of uneaten food. Everything was unwashed and unclean. It was seriously grim.

Paul and I exchanged glances and kept the conversation on the point of our visit as we followed our host up the staircase. Midway up, where the staircase turned, there was a tiny bathroom with the door open. There, standing in all its majesty on the corner of the bath, sat a tub of Vim scourer. The bath was the only clean thing in the house.

It didn't take long to check the roof and we didn't find anything of interest. We made our way back through the kitchen and porch, and went to admire the farmer's herd of Devon Red cattle with him before saying goodbye. They looked in great form and were clearly his pride and joy. Pity the house was in such a shocking state.

Occasionally work took me to Somerset. There I encountered a woman who had committed herself not to a herd of cows but to a collection of rabbits. She lived on her own in a remote spot below the Blackdown Hills, and her ancient house and the adjacent barn were quietly crumbling around her. The house was small, with a narrow, low ceilinged central passage from the front door. The flooring was of bare and uneven flagstones, and as she turned to allow me to follow her inside I noticed, first, that she was wearing wellington boots and secondly, that the floor was strewn with loose bits of hay. She took me through a door on the right into a room that might once have been called the parlour. It also had a low ceiling and was dark and dingy, but the reason for the hay became clear. The room was occupied by at least 20 rabbits. Some were on the floor, quietly pottering about, and others were in an assortment of makeshift cages and boxes placed on stools and chairs and a couple of side tables. Hay spilled out onto the floor and furniture, as did a fair amount of their droppings.

Still wearing her boots, the woman and I made our way up the narrow staircase, over the remnants of a threadbare and faded carpet runner. The rabbits didn't make it up the staircase, but inevitably her boots had been transporting hay and droppings upwards. Her bedroom was a dismal sight. There was no carpet or rugs, just the bare wooden floorboards and a mattress. There were also signs that she was sharing the room with mice and an occasional rat. At least there was a large window, and it was lighter than the parlour below.

There were sloping ceilings and no void to survey, but there was another room at the far end that she pointed to. 'That's where the owl lives,' she said. It had perhaps once been a dressing room or a child's room, only accessed via her bedroom, but it was now derelict. The undersides of the roof slates were exposed and the glazing from the square window frame had long since gone. The pellets and white tell-tale splat marks from tawny owls littered what was left of the floorboards, in among the bat droppings and more rodent droppings. Even swallows had got in and built their mud nests against the roof timbers. It was all a matter of metres from where the lady slept on her mattress.

This was a reclusive existence and a sad picture of decline.

By this stage you may be wondering where the exploding nappy comes into the narrative. The story began when the colleague of a colleague rang to say that a neighbour had given him a dead bat, which the cat had brought in, and did I want it. When you're involved in wildlife this isn't quite as odd as it sounds. The chap was convinced it was a rare species, so yes I did, thank you very much.

A weekend visit was arranged and my husband came along for the ride. We were ushered into the living room and cups of tea were offered while the bat was brought out of the freezer to be examined. It was a rare species – a grey long-eared bat – and the chap had done well to spot this, but it did not take long to confirm. By then my husband and I were sitting in our comfy chairs engaged in conversation with the couple on their sofa. The

Grey long-eared bat

wife was bouncing a cherub of a baby boy on her knees, holding him tall while he practised stretching his legs. He was naked except for his nappy, which looked pretty soggy to me. There was also a slim-built, shaggy-haired dog of uncertain breed who peered intensely at us, trying to get our attention.

Dog hairs began to appear on my trousers where the dog was pressing its chin on my knees, and my eyesight stretched to the carpet where more dog hairs came into focus. Not just a few dog hairs, but an extra carpet of dog hairs. It was a mat of fur. But worse, I realized there were other things on, or in, the carpet as well: bits of cotton thread and … was that a sewing pin? Not just one but numerous pins came into view. Someone in the household liked sewing.

There comes a point when you realize that you're definitely out of your comfort zone. There's a niggling sense of unease and disapproval – who leaves sewing pins lying about when there are babies and dogs present, for goodness sake? – while at the same time you're rationalizing that this isn't your home and everyone has a right to live the way they choose. Best to concentrate on the baby instead, I decided, although the more I looked the more turgid and swollen the nappy appeared. It *really* looked as if it needed changing. And just at that moment, Mum chose to lift the child up by his arms and to swing him round in one perfectly practised movement so that his legs straddled her neck. He beamed with delight and the nappy burst explosively. Urine poured down her neck, drenching her back and her front and the husband got soaked too. She shrieked (well you would, wouldn't you?), so the baby instantly wailed and the dog howled. It was carnage.

We took the rare bat and legged it.

CHAPTER 6

The Lord up a Ladder (Part Two)

By now you'll have realized that this wasn't the first time I'd seen a lord up a ladder. It was several years after the hornet incident and it was a different lord, the late Lord Devon of Powderham Castle. However, the positions were reversed, and this time I was the one looking at a backside.

Lord Devon and the family spent much of their time residing at the castle, but when it was particularly busy with tourists and guided tour groups they took refuge on one of the estate farms. Here at the farmhouse was a normal life away from the public gaze, where garish crayoned drawings from the grandchildren were blu-tacked in pride of place on the kitchen fridge, as in kitchens everywhere.

There were ideas for an extension, and although there was already some knowledge of bats at the property, the specifics of species and numbers and type of roost were unknown. Lord Devon met me on the drive and showed me around, first the barn that was linked to the house and then to the house itself, where we made our way to the first-floor corridor and ceiling hatch.

The next bit can often be a bit awkward. You're in someone else's home but you're the one who needs to enter the roof – and it's not always clear what the owner will want to do. What is the ladder etiquette exactly? Do you gesture for them to go up first since it's their roof, or do you boldly demonstrate professional superiority by grabbing the ladder and taking the first step upwards? You might have done the ladder safety course and have all the health and safety bumf, but you can bet your bottom dollar the owner hasn't.

The etiquette dilemma seemed somewhat accentuated is this situation. I took the bull by the horns and said, 'I'll have a look and see what we've got, yes?'

'Go ahead, go ahead,' he agreed. 'Up you go.'

It was a large space and it wasn't boarded, so it took a bit of time to get from one end to the other. Bats were present, dotted about and hanging quietly and peacefully from the roof timbers, so it was at least relatively easy to do a head count.

Lord Devon was waiting at the bottom of the ladder as I came back down.

'There's a good numbers of bats,' I told him, 'about 17. They're lesser horseshoes.'

'Goodness, how can you tell?'

'Well, they're nice and obliging: they just sort of dangle there like plums. The greater horseshoe is bigger and more pear-sized. It takes a while to get your eye in, and perspective can play tricks on you, but just think pears and plums. It helps.'

Intrigued and enthused, he wanted to have a look. Of course he could; it was his roof. But just to be on the safe side, I added, 'I'd stick to the ladder rather than go all the way in. It's not boarded out.'

So up he went, and then I found myself looking up from the bottom of the ladder. This was a very unfamiliar scenario and I felt a growing sense of panic. Did my professional insurance cover a lord? How long is he going to be up there? What if he slips? Can I catch him if he slips? Of course not. Who'll come off worse if he does? I'll be rubbish at breaking his fall, I'm way too small. I'll be crushed and he'll be crumpled and there'll be broken bones, and oh cripes…

He was saying something, '…quite remarkable…', and his feet were moving downwards. And then he was off the ladder. I quickly popped back up to close the hatch.

'About that offer of a cup of coffee?' I suggested with a great deal of relief. 'Shall we talk this through in more detail in the kitchen?'

CHAPTER 7

The Joy of Ladders

Bat work comes with some serious health and safety considerations. Ladders. scaffolding, rooftops, building sites, underground sites, working in remote locations. The list goes on. Even meeting the roost owner counts as a potential risk on some occasions, although it has to be said it's usually only their dogs that bite. It's the postman's syndrome: we get to the front door and all hell breaks loose. At one swanky house, I'd bowled up and had just rung the doorbell when, at exactly the same time that I dropped my sunglasses and bent over to pick them up, the owner opened the door to an enormous German shepherd dog – that promptly bit me on the bottom.

'Oh!' I cried in surprise, thinking how funny: I've just been bitten on the bottom. And then 'OWW!!' as I realized that it bloody well hurt.

The owner immediately grabbed the dog by the collar and shut it in another room, and after her sustained apologies I went on to do the survey. But it did hurt, especially when I tried to sit down for the next couple of days. It left the most spectacular bruise on my left cheek, which for obvious reasons I couldn't show anyone except my husband.

There were other incidents: a nip on the leg from a tetchy farm dog, some angry ankle-snapping from a psychotic dachshund and one very nasty bulldog type that thrashed the living daylights out of the clipboard that I'd been clutching in my hand. Thank goodness he got the clipboard and not the hand. More commonly, I was faced with dogs that expressed a greater interest in my personal space than seemed polite. Eventually it didn't seem unreasonable to request that dogs were shut away before we entered a house.

In the early years it has to be said that the health and safety aspect of what we did probably wasn't covered quite as thoroughly as it is today. The dangers of asbestos were a regular topic for conference workshops, mind you, and as far as I was concerned learning about something that could kill me trumped anything else that might be on offer, such as learning how to make a felt bat, and which was never going to be something I could list on my annual continuing professional development record.

Asbestos raised its ugly head on various occasions, and it paid to be ultra-cautious – even if the clients weren't. One day I sighted what was almost certainly the classic white, fibrous pipe cladding in the roof of a 1930s house. I promptly stopped where I was, took photographs and went back down the ladder. And then because (on a good day) I'm mostly a conscientious sort of person, I told the owner and then emailed her the details when back at the office, advising her that I believed it to be potentially hazardous. For peace of mind, I also contacted a local asbestos company, sent them the photos and asked their opinion. Difficult to be sure from just the photographs, they replied, but yes, it did appear to be the nasty stuff.

Two weeks later when I returned the woman casually informed me that she'd gone into the roof and scraped off a sample to show someone. My sons tell me there's something called the Darwin Awards for this sort of behaviour.

Relatively recently, I was probably eligible for a nomination myself. I'd been at another old house where I ended up scrabbling about in the under-floor space littered with contaminated rubble dust. My original brief was only to check the roof while works were in progress and I'd been told that the house had been checked for asbestos. But the 'Hey, George, can you look under the floorboards for us?' had been an afterthought by the contractors, and regrettably the asbestos survey hadn't extended into that space. So I'd been down a couple of times before I spotted a dodgy-looking bit of pipe cladding.

It was what you might call a serious exposure, even wearing a face mask, so more than a little concerned I booked a doctor's appointment. He listened, turned to look at his computer screen and then back at me. 'Well at your age, I probably wouldn't worry too much,' he said, smiling. He'd checked my date of birth and was cheerfully telling me that I was so old I'd invariably die of natural causes before the asbestos got me. Don't you just hate it when someone mocks you about your age? It wouldn't have been amusing for someone younger.

When it came to learning about ladders and roof access, the early training was, well, pretty basic, sort of along the lines of 'You can fall off ladders, so try not to, and if it's a really tall ladder, definitely try not to. Oh, and don't put your foot through the ceiling because the owners won't like it.'

These days, who'd have thought there was so much to say about ladders? It's not just about making sure you put them at the right angle. There are portable ladders, step or extendable ladders, different classes of ladders from heavy duty to light duty, construction site ladders and domestic ladders, and so many other aspects to take into consideration such as maximum loading, stabilizer bars and base plates. And when it comes to attics there are the 'fixed ladders', which are a whole family all on their own.

Voluntary visits are now strictly covered by health and safety, and professional ecologists are expected to have done all the ladder courses, the construction site and building courses and all the other courses that shout health and safety, and to have the 'Blue Peter badges' to prove it. I passed one particular course and was issued with an impressive looking card that had my photograph on it, a bit like a driving licence. It boldly stated that I was a 'professional person'. I showed my sons. 'Look! Your mother is a professional person.' I believe they both sniggered.

The majority of homeowners tend not to be quite so health and safety conscious, which is why so many domestic accidents are the result of trying to get in and out of a loft. My feet still quake when I recall a redoubtable old woman of 80-plus and her loft ladder. She lived on her own in a granite-built cottage with small windows and low ceilings. She lived upstairs – the rooms were lighter than downstairs – and there was a small, square ceiling hatch in the centre of the living room, which she said had a fixed ladder.

The situation needed careful assessment, but before I finished saying 'Oh, I don't think that's a good idea…' she had moved a tiny wooden stool into position, grabbed a walking stick and used it to step onto the stool; she then lifted the stick and whacked the hatch so it flopped open to one side. A nylon cord came dangling down from the ladder. She stepped down, flung her arm across my chest and commanded 'STAND BACK!' as she tugged the cord and the ladder shot down with an almighty thwack. It broke every bloody health and safety rule in the book but there she was, still alive at 80-plus. The stool wasn't even solid; it had a tapestry centre.

Many of the fixed ladders we see are plain dangerous simply because they're so old. Positively antique, even. There are ancient wooden contraptions in houses that date from the 1920s and '30s that were almost certainly installed at the time the houses were built. They tend to be incredibly heavy, and sometimes there are weird pulleys and ropes that are needed to bring them forward and unfold them. Then there are some horrible metal inventions, one of which is a concertina ladder that looks like an overgrown piece of Meccano kit. The early versions are ridiculously lightweight so they drop disconcertingly to an almost vertical position (slightly heftier models are still available!). Another design combines cheap metal rods for the foot rests with narrow wooden lengths fixed to the rods, which needless to say often come adrift – so the owner is then left with nothing but the shiny metal rod to step on, which isn't great.

The most outlandish roof access I ever saw was the passenger steps from a small aeroplane. I'd love to be able to tell you which aircraft exactly, but I have no idea; the details weren't something that I would typically note on my survey form. The owner was an enthusiastic plane spotter, and his enthusiasm extended to having acquired the steps and installed them in the roof of his very ordinary semi-detached house. They had shiny, riveted metal panel sides that slid past each other as they unfolded, making for a ridiculously glamorous ascent into the attic. The house was completely unremarkable in all other respects, so the whole thing was very bizarre.

With or without a fixed ladder, the position of the loft hatch can present its own complications as well, and an exasperated 'Why did they put it there?' isn't uncommon. The hatches that have been installed in the high, high ceilings of village halls and Methodist churches seem particularly pointless. Conversely, you find them post-loft conversion where the ceilings have been raised – and the only void that remains is about a foot in height so no one could get in anyway.

Then there are the hatches that have been installed at the top of a vertical wall. You discover that not only is the hatch in the wall of the master bedroom but it's also directly above the king-size bed. So with a great deal of effort you end up having to shift the bed with the owners. This is still preferable to the alternative where one of them announces 'You'll have to put the ladder on the bed.' On one occasion the husband presented a short wooden ladder with a splayed base frame, which he had built specifically to rest on the bed and up against the wall. This property also had a rifle lodged just inside the hatch, so let's just hope the kids never worked out what the steps were for.

Outdoor ladders, the socking big, hefty, double section extendable ladders, are a whole different ball game, and you have to admire those people who can pick them up and throw them around to get them in the right position with what appears to be such ease.

I saw this expertise time and time again on a bat project for Dartmoor National Park Authority. The scheme involved 40 bat boxes that had been erected in the woodland slopes above the River Dart. The initial concept was to use the boxes to survey for the ultra-rare and elusive Bechstein's bat. Studies had shown that these bats feed in the tops of the tree canopy, so the boxes had been specifically erected by professional climbers at great height, all between 6 and 12 metres off the ground. To enable the boxes to be lowered for inspection, they were attached to nylon ropes that were tied near the base of the trees.

Six months after installation there were three of us to check the boxes for the first time. And it immediately became clear that there was a problem. The ropes weren't on pulleys so they didn't feed out smoothly and the boxes twisted and bounced as they came down. Even more alarming, the tie-on points were either directly underneath or very closely underneath each box – which meant that if the rope slipped the falling box would be heading straight for the surveyor below. And these boxes were made of woodcrete, a concrete/woodchip recipe designed to withstand squirrel teeth and frost fracture. This wasn't a potentially nasty bump on the head; it was something potentially much worse.

Urgent phone calls were made. It was agreed that the boxes could be taken down, relieved of their ropes and re-affixed with nails at ladder height. But, and here's the rub, we knew they were intended to be as high as possible on each tree – so that's where the guy with the ladder put them back. The guy with the ladder was Rupert Lane, and he had no fear. If he could get the ladder up and could get to the top of the ladder, that's where the box would go. It just meant that the rest of us would then also have to climb to those heights to check them. The average height was 5.8 metres, with the tallest at 6.7 metres and the lowest at a mere 4.6 metres.

In the first two years of the project the boxes were monitored on ten occasions, after which we switched to checking them just a couple of times each autumn. It normally involved four of us. Rupert, and then later one of the wardens, Rob Taylor, would bring the ladder into the wood on the back of the Land-Rover, and we would take it in turns lugging it through the wood. Several years into the scheme we felt a cluster of 12 of the boxes could be repositioned. These were closest to the river and had rarely shown the same levels of use as the boxes on the higher slopes. They were also almost all at a 6 metre height, and were a pig to reach even with the ladder wielding skills of the likes of Rupert and Rob.

We never did find a Bechstein's bat, but we did find six other species (noctule, brown long-eared, common and soprano pipistrelles, whiskered and Natterer's) and even dormice had moved in towards the end.

If you really, really don't like heights bat work is probably not for you: ladder use comes with its risks and the risks are multiplied from scaffolding. Personally I love getting up high. Whether it's the top of scaffolding or a tall church tower, you get the most wonderful bird's-eye view of the landscape and a bat's-eye view of the roof fabric and the nooks and crannies that they use. As with dodgy ladders, the range between the seriously professional and robust scaffolding and the limp shaky frames put up by some cowboy teams is huge.

The worst example I encountered was at a house on the outskirts of Exeter. The house was built on a field slope, so it was two storeys at the front and effectively three storeys at the back. The builders had been appointed and I needed to attend when they came to remove the roof tiles over the rear wall. I arrived at the specified time and date to find that the scaffolding was shockingly inadequate. It simply didn't have as many poles as it should have had.

In theory there are always a number of options in such situations, one of which is to state categorically that you have no intention of putting your life at risk on such a pathetic construction, that you'll make this clear to the owners and that no works are to start until the whole thing has been built properly. Or you accept the reality that the builders don't care in the slightest if you don't come up as they'll keep working anyway, and the owners don't care because they just want the work done, and it'll invariably be the ecologist's fault if there are delays.

This was early in my career, I hasten to add, before obtaining my 'professional person' Blue Peter badge, and so with a few Hail Marys and a big deep breath, I went up, clinging extra-tightly to the poles that did exist and avoiding the naked edge of the top platform. I watched the roof tiles being removed, confirmed an absence of bats and descended to terra firma, where I immediately went in search of the largest bar of chocolate in the nearest shop.

There are other simple things that you learn about construction sites as the years go by, one of which is to bring your own gear. If asbestos material has been identified, including the cement-bonded fibre roof tiles, you'd normally be expected to wear the protective disposable overalls, the sort you see at the murder scenes in TV crime dramas. If you are relatively short, you don't want the overalls that are in the site office. These will invariably only come in large and extra-large sizes, the crotch will come down below your knees and you'll look like a penguin: all body and no legs.

On some sites you just have to improvise, there are no two ways about it. That's why one day I unexpectedly found myself being sent up to inspect a roof while sitting in a small digger bucket.

It was late autumn, and we were looking at a simple single-storey, timber-frame barn, which had a basic roof of old corrugated metal sheeting. The odd brown long-eared bat, pipistrelle and serotine had been recorded using the roof apex during the summer and it was now time to demolish the building. There are approved ways and means of excluding the animals before you knock down their homes, but these depend on various factors and you can't always assume that they work – in which case you also have to do what we call a soft strip, carefully stripping the roof, checking for bats as you go and before trashing what's left.

The contractors had made a start with some initial ground works and, for reasons that I didn't understand, by digging a trench along the back of the barn. This, for other reasons that were not particularly obvious, had prevented them from putting up the scaffolding. So we had a dodgy roof and we – mostly me – needed to get up close and personal with the tin ridge.

We stood and ummed and erred. The boss was there, and the blokes with the diggers and tools. We debated the use of a hooked roof ladder, and concluded the roof was too dodgy to take the weight. The digger driver suggested using his bucket as a crane. The boss looked horrified, and I could see the risk assessment flashing before his eyes – but time was pressing, and the driver convincingly made a case for putting his mate in the bucket.

'Fine,' said the boss, reluctantly.

The builder was a big burly bloke. Undaunted by the prospect of flouting health and safety rules, he seemed considerably less confident about dealing with a bat if he came across one. We talked it through: 'Keep your eyes open and your gloves on, and check for droppings. Here's a cloth bag. This is how you hold a bat and put it in the bag. 'Pipistrelles,' I said, 'they're tiny, thumb size, you'll be fine…'

And, for good measure, I thought it was also worth pointing out that a tin roof is generally supposed to be rubbish for bats. It has terrible thermal properties so it gets super-hot in the sun and rapidly gets cold as soon as the sun disappears, so in theory he was unlikely to find anything.

The digger was manoeuvred into position and up he went.

Within a matter of minutes he started yelling, 'There's a bat here! I found a bat! It's fucking huge. I'm not touching this.'

So much for the pep talk.

And so there I was, sitting in his place, hovering in the bucket at the side of the roof as I extracted a somewhat surprised serotine bat from the ridge.

'You might as well do the rest of the ridge while you're up there,' the boss shouted.

And there was another bat further along: a little soprano pipistrelle. Clearly these bats hadn't read about the thermal properties of a tin roof.

At the end of the day, it was a job well done. You wouldn't find it in any of the manuals or guidelines, but it was still a job well done. And just to clarify, none of our British bats are really 'fucking huge'. The soprano and common pipistrelles have body lengths of a mere 3.5 to 5 cm and even the serotine is only 7 to 8 cm in length, not much bigger than the size of the palm of my hand.

All light-heartedness aside, there was one particular aspect of the work that we had to take seriously: that was the need to wear gloves, and to educate the roofers and builders to do the same. In 1996, a Daubenton's bat from southern England was found to be infected with a disease called European Lyssavirus, a virus from the rabies family.[1] This sounds scary, but it really needs to be put into perspective. The disease had been carefully monitored in Europe since the 1980s and levels of occurrence were known to be low; but viruses being viruses – and we're all experts on the subject since 2020 – it was inevitable that it might crop up over here eventually. Indeed, there have been other infected Daubenton's bats since 1996, just small numbers, and also a few serotine. The risk to members of the public remains very, very low, first because of such low levels of incidence and secondly because they so rarely come into contact with bats.

The picture is slightly different for anyone who works with the animals, and after 1996 it was strongly advocated – and then formally required – that we should be vaccinated against rabies. My files tell me I was first jabbed in 1997 and have had multiple repeat vaccinations and boosters since then. There's only been

Serotine bat

one death to date, and that was a Scottish bat worker in 2002. He came into contact with a sick Daubenton's bat, and sadly he wasn't vaccinated. I had met him and he was a lovely chap. When briefing roofers on a job, I'd explain this to them: the disease is real and it's out there, but it's not a problem if you're sensible; you don't handle bats unless it's necessary and never without suitable gloves.

CHAPTER 8

Oh Crap!

The American author Bill Bryson is one of my favourite writers, and his account of the history and evolution of the English language in *Mother Tongue* is a great read: learned, erudite and educational.[1] He even manages to get away with an entire chapter on swearing, dedicated to the origins and ever-changing forms of our favourite and most obscure profanities. As such, it seems only fair that I should be allowed to include a full and in-depth chapter on the stuff that comes out of an animal's bottom.

This may variously be referred to as faeces, dung and guano and so on, but most people seem most comfortable when talking about droppings or just plain poo. Maybe that goes back to our early potty-training experiences. I have no idea; it's just a theory.

There are very good reasons for discussing mammal poo. The fact of the matter is it can tell you an awful lot about the animals. Some species, for example otters and other slim-bodied members of the same family such as pine martens, are frustratingly difficult to find and see in the wild, but they frequently and regularly mark their positions by their toilet behaviour. After otters' catastrophic population crash during the 1960s, large-scale national surveys were devised to determine their presence almost entirely by searching river banks and other suitable waterways for their characteristic poo deposits called spraints. And if you really want to get nerdy on the subject, an otter spraint can be distinguished from a mink spraint because it has a pleasant aroma of jasmine tea rather than a … well, a rather pooey whiff. Admittedly, a certain leap of faith is required the first time you get down on your hands and knees so that you can put your nose close enough to find out.

A similar approach was used to try and find evidence of pine martens from their squiggly droppings, in this case aided by advances in DNA analysis to rule out potential imposters such as foxes and stoats.

Poo can also tell you a great deal about behaviour and ecology, and badgers are a great example of this. They usually scrape a small saucer-shaped depression in the ground – we call it a 'dung pit' – and a localized cluster of these pits then become a 'latrine', which marks the territorial boundaries of each social group. If you feed one group with sticky, syrup-covered peanuts to which you have added tiny little yellow plastic beads, the harmless beads pass through the badgers and end up in the dung pits. You feed the neighbouring group with peanuts and a different coloured bead, and voila! Your yellow beaded latrines will define the boundaries of one territory and the other beads will define the boundaries of the other territory. Ecologists get very excited about this sort of thing.

When it comes to surveying for bats in a roof, the reality is that you're actually far more likely to see what the animals have left behind than the animals themselves. Think about it on a temporal plane. Our surveys are so often just momentary snapshots of the roof interior, one or two visits out of 365 days of the year. And bats are mobile. So while we may be lucky and find the bats at home on the day we visit, it's the droppings that we are most likely to see. Hence, it's kind of important that we get our eye in.

The texture of a bat dropping is a good starting point. Because of their insectivorous diet, bats' droppings are basically made up of the small scrunched up fragments of the harder, indigestible bits of the insects. This makes them friable, and if you gently press them between your fingers and thumb they tend to crumble and disintegrate.

Rodents have a different diet and consequently their droppings are rarely crumbly. Those from a mouse will typically look and feel like small black grains of rice: fairly smooth and uniform in size and solid and hard – unless they're exceptionally fresh and a bit squidgy, and then you'll wish you hadn't done the finger and thumb test. Rodents leave other tell-tail signs as well, including brown staining from urine trails and small round holes and burrows in the insulation material, plus gnawing damage.

There are other clues to look for which help build the picture, such as where droppings are located. Certain bats such as the lesser horseshoe bats and long-eared bats favour decent sized roof voids where they can cling to the ridge beams or cluster at the timber junctions, and so the pattern of scatter of droppings often mirrors these features. Many bats like gable wall tops where they're hidden and out of view for the majority of the time, but some of the droppings invariably get stuck to the wall surface. In those locations they almost *have* to be from bats, unless the roof is home to spectacularly athletic mice that can abseil.

Sometimes it's just plain bloody obvious that there's bat poo in the roof, and trainees can get quite expressive about this. 'Blimey, that's a lot of crap.' Or 'Whoa! There's LOADS of the stuff!'

These are perfectly valid observations. However, the lesson is to encourage them to think quantitatively, since eventually they will need to transpose what they've seen into a formal report. 'Loads of crap' doesn't work so well on paper.

Some writers like to quote numbers, which is fine for relatively small amounts, but personally I think it's a little difficult to imagine what 400 or 800 bat droppings look like. Others prefer the volume approach, such as a litre of droppings. Then there's the kitchen method of description, such as a 'spoonful' or 'cupful', which seems eminently more visual and is useful up to a point, beyond which you could well be in bucket territory. And there are still some roof spaces with large colonies, most of which are managed, where we could justifiably describe the quantity of bat droppings by the wheelbarrow load. I'd always wanted to be able to describe a pile of bat droppings as being 'similar in size and appearance to a small molehill', but I've never quite managed to sneak the words into one of my reports.

Having established that bat droppings are indeed present, how do we know which species of bat they come from? Well, obviously when you're training you'll have attended workshops on the identification of bat droppings. Yes, I promise you, there are such things, and some of us even keep our own collection of bat poo for reference and teaching purposes. (Which reminds me of the *Blackadder* sketch when Baldrick is making a cappuccino with extra sprinkles.)

Anyway, it's once again all about getting your eye in, and when you get the hang of it you should at least be able to distinguish between small, medium and large poo. That's not rocket science: it's just that small bats such as pipistrelles make small droppings and big, beefy bats such as serotines make much bigger and coarser-textured droppings. There's a problem when it comes to trying to describe what bat droppings look like because there's very little you can use in the way of comparison. It once occurred to me that a greater horseshoe dropping could be compared to the row of little black seeds in a cardamom pod, but unfortunately that doesn't seem to be an image that readily comes to mind for most people.

Giving students a sphere of reference for deer droppings is much easier. For example, I've always felt that roe deer droppings resemble small peanuts in size and shape (the cheap economy sort, not the posh jumbo sort), while the slightly larger fallow deer droppings are more akin to hazelnuts. On a field trip to Poland one year it was very exciting to come across some freeze-dried elk droppings, which struck me as being remarkably similar to kumquats. And yes, I kept some for reference.

The presence of bat droppings in a roof is certainly a starting point for the assessment that we need to make, and we can build on that by going on to do evening surveys when we watch the bats emerge, or even dawn surveys when we watch them come back to roost in the early hours of the morning. But if we're still not sure what species may be present we can also cheat and get the droppings confirmed by DNA analysis.

We don't get the results back in quite the fantastically speedy manner that we see in the TV crime dramas.

'Have we got the DNA back from the murder weapon yet?' the officer demands.

'Yes, guv,' the dogsbody replies, seemingly less than an hour after discovery of the body. The reality is more like two to three weeks for our results, but obviously that doesn't make for such good telly.

DNA testing of bat droppings is an extremely valuable tool, particularly in some areas of the country where closely related species occur side by side, such as brown long-eared and grey long-eared bats, and where the droppings are indistinguishable by eye. The brown long-eared is dead common and the grey long-eared is ultra-rare. Brown long-eared bats are found all over Devon, and in well-wooded areas they occur in pretty good numbers too, with decent-sized breeding colonies reaching up to 50 animals, occasionally as many as 80. Grey long-eareds, by contrast, have a very limited distribution, which in the southern part of the county mostly just hugs the lowland coastal regions – and they live in much lower numbers too. This means that roosts that are used by a few brown long-eared bats are of relatively low conservation significance because there are a lot of them, whereas roosts used by grey long-eared bats are more important because there are very few indeed.

When DNA results do come back it's jolly nice when they confirm what you kind of suspected but weren't quite sure about – actually it can make you feel positively smug – but it's even better when they tell you you're dealing with something that you weren't expecting and have something out of the ordinary.

Once in a barn loft I found a dozen droppings in a nice discrete location, just enough to indicate that a single, medium-sized bat had been there for a day. We were in brown and grey long-eared territory so I sent the sample off. The results were from a barbastelle bat. This is a widespread but rare woodland specialist and frustratingly elusive. I've come across only 25 individuals or so over the years, and it's been a moment of elation every time.

To test or not to test can be a bone of contention. The DNA testing of bat droppings costs money and it's the owners who end up paying, so it can be darn frustrating when we're required to submit dropping samples for testing even when we really, really do know what we've got.

Sometimes, like so much box-ticking and bureaucracy, it just gets plain silly. I was once asked by NE to provide a DNA test on bat droppings from a long-eared roost near Chagford in Dartmoor for a re-roofing job. The location is important. Three long-eared bats had initially been seen in the roof by the previous ecologists, but unfortunately they hadn't confirmed the species. On the licence application form I simply stated that it was a brown long-eared roost because a) it fell outside the recorded and predicted distribution range for grey long-eared bats; b) there were no grey long-eared bat records in Dartmoor National Park (I checked); and c) I could list a dozen nearby records for brown long-eared bats from my own records, including a 50-strong breeding roost just up the road.

This didn't cut it with the NE officer, who demanded the test to confirm, so with apologies to the owners, back I went. And what was in the roof looking down at me from the rafters? Two lovely Natterer's bats. It was a mixed roost and all the poo was thoroughly contaminated. So on the basis that I simply couldn't give NE the answer it required, I deleted brown long-eared bat from the form and put Natterer's instead. At least I had a cracking photograph to confirm they were definitely present. To her credit, the poor woman who was dealing with the case eventually conceded that perhaps it would be OK to include both brown long-eared and Natterer's on the paperwork. We didn't find either when we finally took the roof off several weeks later.

Once in a while – very rarely, I hasten to add – we come across a roof space that appears remarkably clean. Unnaturally clean, you might say. As if someone might have had a bit of clear up before we arrived but hasn't thought to mention it. It's something we tend to notice though, and it's a bit obvious when they forget to remove the broom. There's a barn loft that comes to mind which had a lovely big snooker table in the middle, covered in a sensible dark-coloured dust sheet. The floor had been swept perfectly and was spotless, but no one had noticed the bat droppings peppering the top of the snooker table.

In addition to bat poo we're always on the lookout for the remains from their feeding activity. When bats catch small insects in their mouths they can generally munch these on the wing, but for those that have chosen larger prey items – big, crunchy dung beetles or juicy, fat-bodied moths – it takes a bit more concentration and mouth manipulation, so they select a sheltered perch and consume the insects there. Several species use favoured roost sites for feeding, and the discarded bits of insect bodies such as beetle wing cases and moth wings accumulate on the ground below them.

We have to be on our guard, though. I've found myself somewhat over-excited on several occasions when I've discovered large concentrations of the wings from orange-underwing moths, only on closer inspection to note a complete absence of bat droppings but the tell-tale white signs of bird droppings. The suspected culprits were probably pied wagtails, and in one case perhaps even a little owl.

There's a particularly well-used greater horseshoe roost 5 miles from where I live, and it's a super little site for trainees to visit. It's a short dead-end tunnel, approximately 40 metres long, built into a limestone slope with a domed brick roof. The owner told me that it was an old venison tunnel, a structure created for cold meat storage pre-refrigeration. The bats can get in from both the original entrance and through a small collapsed section of the roof at the far end, and a range of features have been added to improve conditions for roosting.

It was the beginning of May and I was running a two-day training course on horseshoe bats. I had three women with me from a consultancy from central England – an area without horseshoes – and they were super keen. We had dropped down into the tunnel from the roof opening and were looking at the first point of interest, a very, very fresh pile of greater horseshoe bat droppings and the remains of several recently consumed cockchafer beetles. Not only were there the large, characteristic chestnut brown wing cases of several insects, but there was also one specimen with head, feet and wing cases still all attached. The bat had very neatly nipped off the abdomen – the calorie-important bit – and dropped the rest of the body. Greatly enthused, one of the trainees carefully selected the specimen with a pair of tweezers, popped it in a clear plastic bag and put it in her pocket.

We proceeded to look at the rest of the tunnel and its bats, and after about 20 minutes we made our way out and back to the car. All seated and ready to go, I was just about to turn the engine on when the woman wanted to have another peak at her disembodied cockchafer trophy. She held the bag up to the windscreen for light and simultaneously the legs started moving in mechanical slow-motion. We all screamed like little girls. The animal was definitely dead; it just didn't know it yet. It was really gross.

You might be surprised at the number of other, non-batty animals that also use our roof spaces. The list is quite long and includes mice (house mice and wood mice), voles (mostly bank voles), brown rats, common and pygmy shrews and grey squirrels. The long twisted scats from what are probably polecats or polecat/ferret hybrids also turn up, particularly if the building is adjacent to something like an urban waterway or sewer system where rats, a favourite prey, are abundant. Regionally, the edible dormouse (an introduced species) and the pine marten like to make themselves comfortable in roofs, sometimes with a considerable amount of noise that suggests they are wearing hob-nail boots and have an over-excited social life.

There are also the birds, don't forget. House sparrows, starlings and swifts are well known to use the eaves and roof fabric, but you could easily tot up a list of over 30 species that will nest in buildings in one way or another. There have been a few memorable roof spaces that seem to have a bit of everything: bats, rodents, nesting sparrows, house martins, wasps, hornets, dead flies, spiders. You name it, it's there. This is rarely the case with new houses, though, which by design are intended to be closed and sealed, and hence sterile.

Mice and rats occur in people's attics more than they realize; in fact, it's unusual not to find at least the odd mouse dropping. Rats will predate nesting birds if they can. On my office window sill there are the beautiful long, fine primary wing feathers from a swift, which were all that remained from the bird sitting on its nest at the wall top of a Torquay town house.

Rats seem to make people most twitchy. Even in our work, where we see them relatively regularly at ground level in redundant milking parlours and outbuildings, they still make you jump out of your skin when they suddenly shoot out from nowhere and run across your boot. But hand on heart, I've only screamed once in a roof. It wasn't an old, creepy place but a very average early 1970s block-built bungalow. The roof was L-shaped with three gable end walls, one at each end of the main ridge and a smaller one under the lower and shorter ridge. The roof was clear of boxes and any storage stuff and the insulation material looked relatively clean. I was halfway to the first wall when I noticed a few rat droppings dotted about, but nothing out of the ordinary. Except as I looked, I realized that several of the droppings were fresh. Not old and dried but *fresh* fresh, as in still moist and glistening.

This was unnerving. It's one thing to be out in the open with a rat at your feet; it's another thing entirely to be in a confined space with the animal. 'Come on, Mrs B,' I told myself, 'you do roofs all the time; get on with it.' Talking to yourself helps enormously.

Not finding any bat signs at the first wall, I moved towards the second one and noticed yet another wet rat dropping. There was no doubt about it; there was a rat somewhere very close by. The final bit of the roof was lower, which would mean crawling on my hands and knees. Maybe humming to myself might help calm the nerves. Three metres in there was definitely a noise. I stopped and listened. Nothing. Feeling like an idiot, I started again just as an enormous rat made a dash for it. I screamed as it sprinted past my leg and shot down into the wall cavity.

Fortunately the owners didn't hear me, and I'd regained my composure by the time I'd come back down the ladder. It's hardly going to be reassuring if the professional ecologist is screaming at the wildlife in your roof.

Moving outdoors briefly, another sort of poo that we encounter far too often is of a domestic nature and it's rarely pleasant. We usually have to see what the roof looks like from outside the building as well as indoors, and that means venturing onto the patios and into the gardens – where we find ourselves having to navigate copious amounts of cat and dog poo that has been quietly left to rot. It explains why we need to be so conscientious about our shoes when entering a house; we never quite know what we might have picked up from the previous property.

'So sorry,' the owner says apologetically as they open the back door. 'I haven't had a chance to pick up after the dog for a couple of days.' But when you see the amount of crap that covers the lawn it's blatantly obvious that they *never* pick up. Out of sight is presumably out of mind, but it's difficult to understand just how revolting some gardens get. If it were a public place there'd be an outcry, but it seems to be acceptable if it's outside your own back door.

It can be particularly distressing when you see that children are being exposed to the stuff. It can carry a real health risk from several nasty diseases and parasites, including leptospirosis (Weil's disease), toxocariasis (roundworm) and parvovirus. The most shocking sight was of a tiny child sitting in its nappy on the kitchen floor of a bungalow, a few metres away from the rear French doors. These faced a small area of decking roughly 3 by 2 metres that formed a slightly raised balcony. The decking was completely smothered in a layer of dog faeces, most of it trodden into place by the dog, and I watched as the mother opened the door to allow it to walk back into the kitchen and past the child, carrying goodness only knew what on its feet.

Enough about poo: let's talk about dead bodies instead. Most of the animals that we have been discussing die in the roof space at some time or other. The desiccated corpses of the smaller rodents are most common; those caught with their mouths open at the bait in the snap-back traps or lying next to the poison sachets. The mummified and skeletal remains of rats can be seen too, and once a colleague called out, 'Hey, guys, there's a stinky dead squirrel over here.' That's not something you look forward to hearing.

There have been a few cat corpses along the way as well. Feral cats turn up in barn lofts – maybe they've been eating too many poisoned rats? – but domestic cats will go exploring a house roof if they can smell the rodents and if the loft ladder is down. They've only got to sneak up and in when the owner isn't looking, and then they get shut in. It can be a grim end if the owners don't hear them meowing to be let out.

Bats die in roof spaces too; it's only natural. There's always some mortality among new-born babies and animals within their first year of life. Bad winters and wet summers take their toll, and animals die of starvation – and, exceptionally, what we might even call old age. In well-used roosts where colony animals may be present off and on throughout the year, a few dried bodies are only to be expected.

Bats can also be killed by human objects. Water tanks are a death-trap if they're uncovered, whether water is present or not. If it is, bats presumably approach to drink but are then unable to gain flight lift from the surface. I can think of several large, old-fashioned tanks in a number of properties that were littered with the skeletal remains of bats in the bottom sludge of the water. They turn up in modern plastic tanks too. The tanks come with lids and by law are supposed to be covered, but you'd be surprised how often the lids just aren't there. (You might like to go and check your own.) Other containers with deep sides will also trap bats; common and soprano pipistrelles are particularly susceptible. I've found their small desiccated bodies in a large Quality Street tin, in the deep plastic lid of a cake container and the green plastic contraption used to hold the annual Christmas tree. Very sadly there are often numbers of bats involved, as if the calls from the first trapped animal somehow summoned others to follow.

In the early 1990s I witnessed something very different. I was with a small team of surveyors visiting roosts on the Isle of Man. We'd gone to inspect a church roof in one of the towns where there had once been a record of bat use. The roof was high pitched and tall and some gloomy light came through a round window at the far end. There was little in the way of insulation across the ceiling rafters; instead it was littered with dozens of bodies of brown long-eared bats. We estimated between 35 and 40 of them, scattered across the top of an accumulated layer of old and degraded droppings. The bats were dried and stiff and many were contorted in horrible postures, their feet curled towards their heads, typical of acute poisoning.

This wasn't a natural event. Later testing of several of the bodies confirmed that someone had entered the roof and sprayed it with a timber treatment pesticide containing dieldrin. Remedial timber treatments for woodworm at that time variously contained active ingredients such as dieldrin, lindane and pentachlorophenol (PCP), all of which would eventually be banned for domestic use because of their toxic nature. It was not possible to tell whether the person who had been doing the spraying was doing so to treat for woodworm or to deliberately kill the bats. Either way they must have known the bats were there; they'd have been flying around in panic. It was a terribly cruel and unnecessary act.

Roost owners sometimes find dead bats and, as we've already heard, they generously pass them on for identification or even offer to post them. Yes, I've had dead bats sent in the post. You could say that it's a bit like Christmas when the package arrives, but that would be overstating the anticipation.

There's a proper, sensible way of sending specimens.[2] Not the way one woman posted her bat. She'd heard me give a talk and, finding a fresh body outside her house shortly afterwards, had thought I would like it. She just failed to contact me first. Instead she placed the body between two sheets of kitchen paper, popped it in a brown envelope and posted it. It was August. It was the school holidays and my family and I were camping in Wales. FOR A WEEK. Which meant we arrived back home to be confronted by the most appalling stench in the front porch. In its flimsy envelope, the bat had first been sent through a franking machine, and then the pressed remains had sat decomposing on the door mat in the summer heat.

The package had to be disposed of outdoors and promptly. Husband and sons were seriously unamused. So was I, but it seemed only proper to check the contents first – so with a scarf wrapped round my face and with gloves and a grimace it was taken to the furthest end of the garden. Most of the body had oozed into the kitchen paper, but a thumb bone could be seen and there was just enough of an imprint of one ear membrane to be able to identify it as a long-eared bat. It got a decent burial near the compost bin.

The owner had thoughtfully put her name and address on the back of the envelope. And did I write back? Damn right I did, although not very politely. I was in the dog house with the family for ages, *and* we had to get a new door mat.

CHAPTER 9

Can I Help You, Officer?

If you're on the night shift for your bat work, it's sort of inevitable that sooner or later your activities are going to be misconstrued. I learnt this fairly early on at a house on the outskirts of Honiton. It was one of three houses that were nestled within their respective gardens off a short, shared drive from the main road. There were no gates, but each of the properties was reasonably well bordered by thick leafy hedges and some decent-sized overhanging trees. My colleague James and I had bowled up just before sunset in time for the evening survey. The owner was out but knew of our visit.

The drive was tight, and we'd left the car just off the main road so that we didn't block access for the other owners. It was a noisy march over the gravel, then we paused and conferred over who was to stand where. James took the garden path and went around the side of the house so he could cover the rear and I chose the front lawn, just off the drive.

It was a slow, quiet start, and then bats began to appear from somewhere behind me, not from the house we were meant to be looking at but from one of the others. We're supposed to be reactive when this sort of thing happens, so there's a bit of dashing about, ducking up and down to get a better sense of direction, and instinctively that's what I started doing. The bats' behaviour was confusing and it wasn't at all clear which house they were coming from.

'GO AWAY!' a man shouted.

The voice seemed to come from behind a thick laurel hedge.

'Hello,' I called in reply. 'Sorry to disturb you; we're doing a bat survey.'

'GO AWAY!' he shouted again.

It didn't seem as if he'd heard me, so I repeated myself. Nothing happened.

'I KNOW YOU'RE THERE!'

'Yes, I know, I'm sorry, we're doing a bat survey.' Again, nothing.

'I'll CALL THE POLICE. I WILL, I'LL CALL THE POLICE.'

I tried a few more times, but he sounded elderly and anxious, not to mention angry, and my disembodied voice was obviously not helping matters. It might be possible to find my way round to his side of the hedge, where I might give him a heart attack or he might have a gun (you never know), or I could yell to James that we needed to go ASAP.

'Sorry,' I panted, as James and I trotted down the drive. 'I should have asked the owner to let the neighbours know.'

The second time this sort of thing happened (at a different property), I fully admit that I was inadvertently trespassing. In my defence, the owners had failed to give any indication that the large green expanse of lawn that lay directly in front of their cottage didn't belong to them. It was only separated by a very low stone wall, no more than a foot high, which also had a flagstone path through the middle of it, so I had simply assumed the one was connected to the other. Not so.

There were three of us, and we had come to do an evening survey of the cottage. It sat at the eastern edge of a small cosy quad of properties with the lawn in the middle. To the north was a large, imposing house that lay sheltered and screened by a taller wall covered in fancy trellising and roses, a rather lovely aspect that made me think of Jane Austen dramas. There were trees to the western edge of the lawn and a stone-flagged drive and barn conversion to the south.

Our cottage was full of brown long-eared bats and we wanted to know where they all emerged. Andy, one of my surveyors, went round to the back garden and the other, Ruth, took the north end. I opted for a position off the south-west end where I could cover the front. The evening gradually darkened, and as it did so the long-eareds began to appear and flutter past. They were materializing from somewhere in the centre of the front wall, beyond my direct line of sight, so I needed to move – and the best view was from the lawn.

Long-eared bats can be challenging so you often need to cross-reference with your colleagues. Ruth and I started to call: 'Are you seeing them too?' 'I'm up to 17.' 'That one went your way.'

Suddenly out of the darkness I saw a man striding across the grass in my direction. He was silver-haired and distinguished-looking, and he wasn't smiling. Most surprising was the fact that he was wearing an outrageously ostentatious satin kimono, full length, vivid blue and enlivened with golden swirling dragons.

'What the hell do you think you're doing?' he barked.

As you may have realized by now, conflict situations aren't uncommon in our work and I've dealt with a number. I've run workshops on the subject, for goodness sake, but I'd never had to appease a man wearing a kimono in the middle of the night.

His tone was angry and imperious. I was stumped for words. He looked absurdly theatrical and I still really, *really* needed to keep my eye on the bats that were coming out.

'What are you doing on my lawn?' he demanded again.

Rules of conflict management demand that when you're caught with your trousers down you look the person in the eye and apologize. Which is difficult to achieve when you have your eyes glued to activity in the *opposite* direction of that person. So, not looking at the man and his kimono, I earnestly apologized and explained the error, while slowly edging towards the hard flagstones of the cottage.

Brown long-eared bat

He wasn't going to be mollified by this, but Ruth called out at this point and, perhaps sensing he was about to be outnumbered, he gave a disgruntled harrumph, about-turned and marched back to his house.

'What did Nanki-Poo want?' she asked.

'I was on his lawn. I did say sorry, but guess what, I recognized his face and he's an actor on the telly.' And, while still counting the last of the bats, I described his role in a particularly well-known TV series.

On another occasion, it had just gone one-thirty in the morning and we had been parked by the side of the lane for the last 40 minutes. We were on a narrow pull-in which during daylight hours would have given a nice view of the sloping fields down towards the fishing town of Brixham and the blue seas of Torbay. But it was pitch dark, there were no street lights and not a single car had passed.

I'd finished a rather soggy sandwich and was on the last dregs of my coffee. Mark Robinson, an experienced field worker familiar with the long hours of boredom in between bouts of frantic and manic activity, was working his way through a variety of activity foods and had just produced a paper bag of fresh cherries.

The greater horseshoe bat that he'd been radio-tracking was from Berry Head, a cave roost in a redundant quarry, carved out of the 60 metre tall limestone headland to the south of Brixham.[1] It had spent the first part of the night bombing along the field edges and zig-zagging around scrubby slopes, stuffing its face full of dung beetles and chunky moths. It – she – had not been pregnant that summer so there was no baby and no need for her to wing it back to the maternity roost to suckle. Instead she had nipped into an abandoned and derelict coach house so that she could quietly fart and poop while digesting what was in her stomach. She wasn't in any great hurry to go anywhere and all we could do was wait. She might set off for another hefty feeding bout, and if so we wanted to know where, or she might take a straight line back to the roost. (Females that have a pup to suckle can feed so intensely during these feeding bouts that their stomachs become distended, and they can be mistaken for still being pregnant!)

Mark and I had been chatting to pass the time, as you do. Kids and family, background, friends and colleagues in common, all that sort of stuff. I think we had just got to the part where he was talking about playing the drums, or it might have been a guitar; that bit is rather vague.

Suddenly there were car headlights coming in our direction. It was a police car. It slowed down, passed us and then reversed so that it was level. The driver door opened, and Mark too was out like a shot.

'Morning,' I heard him say. 'Would you like a cherry?'

And he proffered the paper bag to the policeman, who smiled and said, 'We thought you might not be wearing any clothes.'

The second policeman came round the front of their car and I thought it best to join in, so there we were chatting, at what was apparently a favourite local spot for couples who park up and take their clothes off. It must have been fairly obvious we weren't up to anything, I was seriously togged up to keep the cold out and Mark was well into his spiel about the bat survey project. And he *had* notified the local Dartmouth police station in advance, he said. It seemed that we were more of an interesting diversion than a situation of unlawful intent.

They eventually drove off into the dark and we got back into the car.

'Bastards,' Mark muttered. 'They ate all the cherries.'

Six years later, and I was in another car with my friend and colleague Paul Chanin, and we had pulled over to a layby. It was five in the morning and it was just getting light. It was the end of a very, very long survey session that had started the evening before. We were covering the Plymouth site again – not the buildings, we'd done those, including the infamous farmhouse with the tub of Vim – but the broader countryside. In those days such large areas were typically surveyed by a combination of methods including 'walked transects', where some poor sod wandered about the fields and woods for hours clutching a bat detector; 'static recordings', where a detector might be strapped to a tree or gate post for a week or so; and 'car transects', which is what we'd been doing. It basically involves driving very slowly along the meandering lanes with a bat detector and an orange hazard light fixed to the car. It's crude by today's standards and the results are seriously biased towards the loud shouty bats that fly in the open, but it can form a useful part of the overall approach.

Paul and I had done the evening shift from sunset, had taken a coffee break at around one o'clock and had then done another three and a bit hours until dawn. We'd pulled over, brought the gear back into the car and were sitting unplugging cables and finishing the time sheet. We were both dead tired, really, really tired, and all we wanted to do was to head back to our respective homes.

The police car appeared out of nowhere and stopped in front of us.

'Oh no, not again,' I thought.

Remembering Mark's example, I got out of the car to (a) look friendly and (b) demonstrate that, yes, I was fully dressed. The police, a man and a woman, were already walking over.

'Any problems?' the woman PC asked.

'Hi there,' I said. 'No, not at all. We're ecologists. We've just finished a bat survey and we're packing up the equipment.'

'Oh wow!' she exclaimed. 'That's brilliant! That sounds so interesting!'

She turned to her companion. 'Wasn't I talking about this only the other day, Phil?'

Phil nodded faintly in a non-committal manner.

She carried on. 'Do you know anything about bat boxes? I'd really, really like to put some bat boxes up at our place. What do you think?'

What did I think? I wanted to get back home and go to bed, that's what I thought. But hey, let's not get surly with the police. Twenty minutes later she'd extracted as much information on the subject of bat boxes as she wanted and we were allowed to head off home. Her interrogation methods were superb.

CHAPTER 10

Batting from a Boat

By this stage of the book you probably won't be surprised to hear that there have been evenings when we've carried out our bat work from the water. Sometimes you just have to think outside the box to try and find the answers that you want. The answers to questions such as:

a) When all the bats in a particular roost leave each night, which routes do they take to fly into the broader countryside to feed? Do they rely upon some routes more than others and are any of these routes vulnerable to obstruction?
b) The bats on the east side of the River Dart turn up on the west side of the river. Are they taking the long route round, flying miles upstream and using the nearest bridge, or are they actually crossing the river, and if so, where?

How you answer these questions depends on manpower, funding and the technical equipment that's available to you at the time. Today the technology for following and plotting bats in flight is super-sexy, ultra-sophisticated and all-singing, all-dancing, but it hasn't always been thus.

In the first instance, Peter Chapman suggested that I should spend my evening sitting in a small boat on the River Dart in 1990. Peter, if you remember, was the warden to an important roost of greater horseshoe bats that was owned and managed by the VWT.[1] The bats were present all year. In the summer months they bred in high numbers in their bespoke five-star barn and in the winter they hibernated in the old quarry caves behind the barn. It was, and probably still is, the largest roost of its sort in Britain and possibly in Europe, with numbers in the summer rising from fewer than 500 bats in the late 1980s to well over 2,000 in more recent times.[2]

The site is located in the small town of Buckfastleigh on the south-east edge of Dartmoor. The River Mardle runs off the moors and passes through the town in an easterly direction, while the River Dart rushes past Buckfast Abbey from the north as it heads south towards Totnes and Dartmouth. Both rivers pass beneath the A38 dual carriageway and then join on the east side of the town. From previous observations Peter had seen that the majority of the bats favoured high-tailing it out of the quarry along a dark, sunken footpath to the north, avoiding the majority of the roads and streetlights. He also knew it was more complicated than this and that they must be using other routes out of the town before fanning into the countryside. He wanted to get a better picture, including what was happening at the river junction just below the weir.

He'd asked members of the bat group for help, and we'd bowled up at the beginning of the evening to be given our instructions. I'm pretty sure the boat hadn't been mentioned. Nevertheless, 20 minutes later there we were, me and this other chap, both having been expecting an easy evening bat watching, clutching at the hazel branches as we slid down the river bank towards a small inflatable. We scrambled in. Peter fed the rope out for about 5 metres, retied it to the tree and then left us bobbing in the channel, gently swinging side to side in the current.

It was past sunset and the light seemed to fade much more quickly on the river, shadowed by the overhanging tree branches and the blackened surface of the water. We sat in the dark feeling like right chumps. It took a while to adjust to the noise of the water from the weir and cars passing on the road on the far side. After at least 20 minutes our bat detectors began to shout at us, a muffled mix of calls from pipistrelles and other species that were harder to confirm, and then the warbled fragments from greater horseshoe bats could be heard as they streamed past. We couldn't see much (no night vision equipment on this occasion), but we could get a rough idea of numbers and a strong sense of direction as the horseshoe bats sped south down the river.

The stream of movement didn't last long, although it felt as if we'd been on the water for an awfully long time before Peter came back to pull us in. He was pleased with what we reported: confirmation of another well-defined commuting route for the colony. I think I'd have been a tad happier with slightly less darkness. We could have had an entire otter family circling the boat in curiosity and we'd never have known.

Sixteen years later, I was asked to sit in another boat on the River Dart, and it was a rather more scenic experience. We were further south where the river is tidal, where it is overlooked from the west by the historic buildings of Dartmouth Royal Naval College and from the east by the old steam railway line. It is a truly, truly beautiful stretch of river. It twists and winds and broadens and narrows, all against the backdrop of sloping fields, wooded gullies and shallow muddy creeks. Kayakers love it. Ospreys stop to feed for weeks in the late summer and early autumn before migrating. Agatha Christie chose to live here at Greenway House on the opposite bank from the small, picturesque hamlet of Dittisham.

I was helping out on a project organized by Steve Markham, and we were again trying to piece together the movements of greater horseshoe bats from the Brixham caves. In the past, the conventional wisdom had generally been that these large, low-flying animals would simply not cross open treeless spaces and that wide rivers would effectively act as barriers, but this was now being questioned. The radio-tracking study carried out by Mark Robinson and his co-worker Maurice Webber had strongly indicated that the bats *were* crossing the river, or at least a few of them were,[3] but they just hadn't had the chance to prove it.

Steve wanted to find out for sure, so he needed to think like a greater horseshoe bat: 'Where would I most like to cross the river?' The logical answer was places at which the most favourable contours led to pinch-points; in other words, where the river was at its narrowest. He went through all the data (the direction in which bats had been recorded travelling, all the roost records, his 3D contour maps and landscape maps) and there were indeed a couple of promising possibilities – firstly, between the marina on the east bank and the opposite shore just north of the Higher Ferry, and secondly, at the Greenway/Dittisham bend. He decided to put two of us in a boat.

A small fishing craft and skipper were hired from Kingswear. We aimed to pootle upstream to be off Greenway House close to sunset time, which meant that we got to enjoy the sun as it set over one of the loveliest rivers in the country and to see herons, little egrets and kingfishers. And even when we were loitering on the water after the sun had gone down, there were the silhouettes of the sailing boats and the genteel slopping sounds from their moorings.

It took three evenings, seven hours in total, before we both simultaneously heard the unmistakable calls of a passing greater horseshoe bat from our detectors. We were bang-smack in the middle of the channel, quietly bobbing up and down as the bat flew low, close to the water, heading in a straight line east towards Greenway Point. There was no doubt about it, we'd clocked one. It was brilliant. And then just to prove the point there was a fainter warble of a second bat, not as well defined but clearly flying over the water ahead of us in the same direction.

We can use all sorts of fancy recording equipment these days, but there are still times when actually being there and seeing what's what is what counts. When you look at the maps and theorize, rivers look like great, wide, empty expanses. But they don't have to be. Our little stretch of the River Dart was dotted with moored sailing boats, good solid objects providing a textured landscape within which a bat could fly and navigate.

CHAPTER 11

The Jacket on the Back of the Door

A common question in my career has been 'Don't you mind working in the dark?' Or a variation on the theme: 'Isn't the work a bit spooky?' The first question is the easiest to answer; the second is much harder and depends entirely on one's view of the human condition and whether or not you believe in ghosts.

To begin with, lots of people work at night and it's not always in the dark so it's not really a big deal or worth over-dramatizing. And it should also be clear by now that the vast majority of night-time bat work is carried out with colleagues. It usually takes more than one person to do an emergence survey of a building. There's no point only covering the front of the house if the bats can nip out the back, sniggering as they go without having been seen. And then there are some locations where it's just plain common sense to make sure you've got a buddy. It might be rural and remote, in which case don't go on your own! Or it might be an urban and decidedly dodgy neighbourhood, in which case *really* don't go on your own!

If you're doing an evening survey in town you'll probably have to stand in one spot on the pavement for quite a while, which could easily be misinterpreted. Definitely wear a high-vis jacket so that you look official, and definitely be within shouting distance of your mate. The drunks can be a distraction but have always been harmless in my experience.

I'm less keen on the surveys where we have to walk through the countryside for several hours at night, what we call walked transects. It'll usually be as part of a team but often with team members taking different routes: the north side of the wood while you take the south side or the east end of the field valley while you take the west end. Bright moonlit nights are nice and atmospheric in the damp dew. Dark woodlands can be more disconcerting, particularly if a daft, short-sighted badger suddenly trundles onto the track directly in front of you or a roe deer explodes noisily from the undergrowth. It'll certainly get your heart pounding.

If you're looking for the positives of bat work, and a lot of ecology field work in general, much of it is pretty physical and could, arguably, be good for one's physical well-being (ignoring the dodgy ladders and shaky scaffolding of course).

In 2015 we covered a particularly large site where greater horseshoe bats were a key consideration. This meant that we had to do a total of ten evenings during the year from April through to October, each survey lasting a period of three hours or more. The site was split into three transect routes with each surveyor walking roughly 5 miles a night. The maths are easy: if you walked all ten transects you'd have walked 50 miles that season. There were also the extra miles it took to install and collect several static recorders, *and* we were doing multiple daytime bird surveys and inspecting dormouse tubes and reptile mats. I'd probably done the best part of 150 miles by the end of the project.

What about derelict and empty buildings? There have been so many: farmhouses and manors, barns and milking parlours, hospital buildings of all sorts; even mortuaries as well as churches, chapels and warehouses, Napoleonic fortifications and German bunkers. Most have been structurally safe and accessible, or mostly safe; others definitely unsafe. That doesn't make them scary; it just means they need to be treated with respect.

Admittedly, while these sorts of buildings may not be structurally unsafe they can be pretty unpleasant. There's one that comes to mind, an unlit, disused underground car park near the harbour in Torquay that I refused to survey without a minder. It had been taken over by homeless sleepers and drug users and was utterly squalid, strewn with filthy, tatty sleeping bags, empty booze bottles and tins, human excrement, needles and condoms. Even my minder, a short bulky security guard, didn't want to loiter there. Unsurprisingly, there were no signs of bat use.

Are these places spooky? No, for me spooky is the wrong word. I looked it up and it means 'sinister or ghostly in a way that causes fear and unease'. None of the buildings I've surveyed have ever felt spooky. A bat flying out of the dark and up and down an empty corridor has only ever been wildlife doing what it does, and it's never been anything to be afraid of. And I can't say I've ever worried too much about bumping into some mad axe-killer psychopath loitering in such places either. That's mostly the stuff of horror films.

Nor do I believe in ghosts. But … and there is always a but … there have been times where the human history of a place has jumped out at us, not so much the ghosts but the whispers from the lives that have been and gone. These places have proved to be unsettling and surprisingly moving.

The buildings of one redundant hospital were particularly troubling. In the old maternity ward, where new lives had been born, there was mould and decay, broken glass, doors off hinges and rotten notice boards. Bats had moved in through holes in the windows and were using some of the smaller store rooms, but it somehow didn't feel right that they should be there. In the mortuary wing, in the room where the deceased were brought to lie in rest, the ceiling had partially collapsed and the tiled floor was cracked and half-bare. Whatever respect there may once have been for the dead in this room, there was nothing to show for it now.

In the mid-1990s I visited a crumbling old farmhouse, a former manor house from the early eighteenth century. It was in a shocking state: it had been neglected, added to and neglected some more and had been largely unoccupied for decades. I've thought of it often over the years, and one of the bedrooms still remains poignant.

You entered the house through the large front door into a wide hallway, with a door to the kitchen on the left and a room to the right. A broad staircase, which was partially collapsed, curved up and round to the first-floor corridor and rooms off to the front and rear. At the back of the kitchen was a door to a much smaller and steeper spiral staircase, the one that servants would once have used to reach the second floor, and their dark attic rooms. At some point an extra wing had been added to the north-east side of the house, but the roof was incomplete when we visited.

The 80-year old farmer, who had recently died, had spent his last years living mostly in the kitchen, with a sleeping bag in the front room. The sleeping bag had not been moved. We looked in the kitchen. It was dark and stark with a bare flagstone floor, a partially boarded window and a sheet of naked corrugated tin that stood erect where the back door had once been. Ivy had just started to creep over the top. There were only remnants of the farmer's occupancy – a small table and single chair and some tatty crockery by the sink. It was a shock to see a metal gin-trap hanging on the wall. These ugly, inhumane traps had been banned since 1958, almost 40 years earlier. Yet there was one still on the wall.

We gingerly navigated the main staircase. The door to the first front-facing bedroom looked normal. Empty, but normal as far as empty rooms go. Then we opened the door midway along the passage to another room. It had two sash windows and was surprisingly light, even though buddleia had taken hold from the outside guttering and had begun to stretch branches through one of the windows. The room contained three metal frame beds, each with a bare mattress and a thin grey pillow, and a large enamelled washing bowl sat on a wooden stand. On the floor were several newspapers dating from 1945 or 1946, I can no longer remember which exactly. As the door swung gently back behind us there was the khaki jacket of a British soldier still hanging on the hook, almost clean and fresh-looking.

It seemed so casual and so intimate, as if the hand that had put it there had only just done so. The house had been used as a Second World War hospital. That room had remained unchanged since those last soldiers had stood up from their beds and walked out the door. It wasn't so much the whisper of a life that had been and gone but something rather louder. Bats were living in the attic rooms and in the roof of the north-east wing, but somehow, just for once, their presence seemed less pertinent.

CHAPTER 12

Call the Press!

In 2006 I found myself walking into an airing cupboard with the actor, comedian and TV presenter Bill Oddie. I think I'll come back to this in a minute.

Animals make for good media stories, a lesson learnt early as a young zoo keeper. If you can throw in a conservation twist even better, although the cynic in me began to harbour a strong suspicion that the animal was always the main attraction for journalists in the 1980s – to hell with the background message of population decline or persecution. Regardless of species, any new birth was considered a photo opportunity and the press office would be quick to jump. The general approach (a clear sign of the times) was to grab a female keeper who could pose with a small cute and cuddly animal and beam with delight, while the blokes, who made up most of the keeping staff in those days, were the ones who were told to look on in a proud, manly fashion at the tiger cubs and the bison calves, the young of the dangerous and heftier species. Thus there is a rather cringe-worthy photograph of me aged 19 clutching a baby Nigerian dwarf goat, which my mother had on her wall for years. There are others taken the following year of me bottle-feeding a baby Asian short-clawed otter, and the year after that yet another, this time with a baby white-faced Saki monkey, also being hand-reared. Clearly there's a bit of a theme going on here.

The press like a good bat story too, but the positive or negative tone of an article and the level of factual accuracy can vary beyond belief. I have to say it's been a love/hate relationship over the years.

One day, during the period when the Bat Project was based in the Baltic Exchange Building in London, we received a call from the local branch of the Midland Bank to say they had a bat roosting on their outside wall above the cash machine. This was in the City, a five minute walk from St Paul's Cathedral and not the sort of place where you expect to see a bat, let alone one roosting in broad daylight on posh concrete. It had definitely taken a wrong turn somewhere and was almost certainly going to be exhausted, so I was sent to collect it and bring it back to the office. Tony Hutson peered at it intently, measured it and pronounced that it was a Nathusius' pipistrelle. Remember that at this time the species was still classed as a rare vagrant, one that had seldom been recorded in the UK except in a few disparate counties, the Channel Islands and on the odd North Sea oil rig. It was the first I'd ever seen.

The bat was passed to another expert for confirmation, given water and a snack and released the following evening near Hampstead Heath. It made for a *really* interesting story, and remarkably some of the detail did make it into the newspapers in a moderately sensible form. The headlines were awful, though, based almost entirely on jokey quips about the bat wanting to make a cash withdrawal.

This should have warned me of the pitfalls of media coverage to come. The office move to Paignton Zoo came with the added advantage of the zoo's media links and we certainly got coverage, but blimey, it was often depressingly unoriginal, dominated by Batman and Dracula references and tardy on the fine detail.

A piece in the *Herald Express* on 7 December 1989 read 'Jokers had better beware – here come the batmen!' followed by the introduction 'Britain's batmen have moved into a new bat cave – at Paignton Zoo.' Nope, it was just me, and I'm not a bloke.

Perversely, I did like the cartoon: Batman and Robin in classic TV outfits standing at the entrance turnstile of the zoo with the caption 'OK, so where's our new office?'

Two other pieces came out a couple of months later. The first exclaimed 'By George, she's batty about bats' and included a photograph of 'Batwoman – George Bemment', an attractive, smiley blonde who very definitely wasn't me. God knows who she was. The same newspaper later gave a short column to advertise a talk with the weird headline 'Bad Press Bat Squeak' and the bright, cheery introduction: 'BATS – victims of a bad press over the years thanks to the likes of Bram Stoker – are all ready to bite back, thanks to the Paignton Zoo-based Bat Project.'

It was like, '*Really*?? Is this the best you can do??' but there was no fighting it. We just had to smile, take it on the chin and persevere. We knew we were making progress; the Batman and Dracula associations would gradually fade and bats would eventually join the mainstream of natural history. Just like sharks!

Newspaper cuttings from 1989 to 2007 from an old office folder suggest the tone was shifting. There are the well-defined bat stories such as rare and unusual finds: an account of the exhausted and emaciated Leisler's bat that had been found on the primary school playground and an unexpected grey long-eared found at the entrance to a Torquay secondary school. The first of these included a photograph of a Leisler's bat, credited to the professional photographer Frank Greenaway. The second article also included the photo of a grey long-eared bat but failed to credit the photographer as it should have done. It happens to be another one by Frank, one that appears in the book *A Field Guide to British Bats*.[1]

Two years later a second grey long-eared was found at the same school, this time in the bike shed. (The students who discovered it had obviously gone to have a quick puff: the butt ends on the ground were a bit of a give-away.) Such a repeat find in such a location made for another news item, but this time the photograph was of the bat in my hand, taken before it was released.

There are the human stories that just happen to include bats: articles on businesses and companies that donated bat boxes for the Devon Bat Group; a story about an owner who had decided to build himself a little hibernation tunnel for his greater horseshoe bats at a cost of £10,000; and a master thatcher who constructed a beautiful dormer entrance into a farmhouse roof, the only such feature of its sort in the south-west as far as we knew. All the photographs are of the people, smiling at the camera while clutching said bat boxes or theatrically clutching an armful of reeds on the thatched rooftop. All positive, feel-good stories, and not a vampire in sight.

I particularly like a piece from September 1992 reporting the finding of a stray Daubenton's bat on the outside wall of a building overlooking Torquay harbour. It was the wall of the local branch of the Midland Bank. What are the chances?!

If the work requires media coverage, radio interviews and being filmed are all part of the same game. Radio was generally fine, but the first time in front of a camera was a highly uncomfortable experience. I had not been on the Bat Project for long and I was nervous. To compound matters, the instructions were to get to the Sky News studio at 6.30 am for make-up. But I was only going to get a five-minute slot, so why the need for make-up? Studio lighting, they explained. I expected a bit of lippy and perhaps some extra mascara, not a 20 minute operation to apply layers and layers of different creams and powders. This wasn't me. Where the hell had my freckles gone? I felt like a pantomime dame.

It was very difficult to feel relaxed, what with the strange face, the alien environment and, worse, the artificiality and triviality of the studio dialogue. I suspect that all the right words probably came out ('timber treatment, poisonous, new laws'), but it must have sounded awkward. There would be other interviews on camera in the years to follow, but all in fields or in the aforementioned airing cupboard, and that was all much better as far as I was concerned.

The smaller local TV companies and news crews face a problem when they want to show footage of British bats. We're not talking about the professional planning, heroic patience, technical wizardry and astronomical budgets of the natural history film-making industry; we're usually talking about a short-notice, click-of-the-fingers, 'how-quickly can you be available?' request on a tight budget, during the day. And as we know, bat roosts can be physically challenging to enter, and if there's going to be disturbance the filming might need to done under licence. You could film when the bats aren't there, but what's the point of that?

Greater horseshoe bats are high profile in Devon. They're big, they're chunky and if you can say that a bat has charisma then a greater horseshoe has it. And the population recovery within the county over the last 40 years makes for great feel-good TV. But where do you film? The answer is in a field full of cow dung, preferably on a slight slope with the Torbay sea in the background. We've all done it: turned up with brushed hair and clean clothes wearing a pair of wellington boots. It's all to do with the bat's diet and a preference for dung beetles. There's a knack to sounding enthusiastic about cowpats, and sometimes you get to poke around to look for grubs just for good effect.

And did you know that it's possible to draw a link between the production of organic apple juice and greater horseshoe bat conservation? I agree, the link isn't immediately obvious, perhaps a bit of a three o'clock in the morning marketing idea, as Eddie Izzard might say, but hear me out. The company producing the apple juice, along with a rather delicious range of other fruit drinks, lemonades and ginger beers, was based in south Devon not far from the River Dart, an area that, as we've already heard, is heaving with greater horseshoe bats, relatively speaking. The organic apples get crushed for the juice and then the pulp goes to a nearby farm to feed the organic dairy cows. Voila! Healthy, chemical-free (fruity-flavoured?) cow pats, ideal for dung beetles and greater horseshoe bat take-aways.

The interview was particularly jolly. The sun was shining brightly after days of rain, and although the field was horribly soggy that was fine; I had come prepared with my wellies. The cameraman and the woman producer didn't have any wellies, but being true professionals they tiptoed and stumbled and sloshed their way into the middle of the field to get the best background view. And there we stood, knee-deep in mud and dung in the name of bat conservation and a clever product sales pitch. I didn't get paid, I'd like you to know, although I was given several boxes of fruit juice, which was delicious.

When we shot a TV piece in those early years we rarely knew when it might air, so the chances are we'd never see it – but it was worth keeping an eye open. The TV channel might have gone to their film library for a five-second sequence of a bat in flight just for a bit of action. If the subject was greater horseshoes, then with a bit of luck the bat they showed might also have been a greater horseshoe… but it didn't always follow. Sometimes they clearly thought any old bat would do, even a fruit bat, so once again we were left helplessly groaning at the telly.

And so to the airing cupboard. This was for a short BBC series called *Bill Oddie's How to Watch Wildlife*, specifically the episode entitled 'Homes and Gardens'. The producer was interested in a couple of 'ordinary' homes where the owners cheerfully lived side by side with their bats and where it would be possible to film the animals in situ. We came up with a couple of potentially suitable properties. The first was a very average 1960s chalet bungalow, tucked away on the outskirts of Kingsbridge and occupied by an elderly lady and a nice little colony of lesser horseshoe bats. The other was not quite so ordinary, an unoccupied listed building that was looked after by the caretaker farmer who lived in the farmhouse across the yard; it was a good user-friendly site where we could usually expect to find small numbers of bats of several species. Students loved it.

The producer, cameraman/sound recorder, Bill Oddie and I arranged to visit the bungalow first. The garage was attached at one end and had a large, pull-up metal door that had been jammed in the open position for as long as I had been visiting. The owners, including the husband when he had been alive, had been reluctant to fix it since it was the bats' route in and out. An open ceiling hatch allowed the bats into the garage roof and from there through a small hole in the inner block wall into the roof of the bungalow. If *we* wanted to get into the roof we had to go through the back of the airing cupboard.

The priority was to minimize the amount of time in the roost, and the early morning was the best time to film. The sequence that appears in the series shows a snapshot view of the bungalow then me and Bill walking up the drive, climbing the staircase, entering the airing cupboard and then crouching in the roost and briefly quietly chatting. In fact the entire sequence was filmed in reverse. For quite some time we organized ourselves outside the airing cupboard: sound check, wires, an explanation of where the bats might be positioned, what we could and couldn't do and how long we would have. We then went downstairs, where we were filmed walking back up the staircase again. And then we went outside, where we were filmed walking towards the property.

For a non-TV person the filming process was interesting although slightly disappointing. The entire shoot, including the visit to the second building, took just over half a day, yet the sequence on film lasted less than five minutes. Filming out of sequence was a surprise (how naïve, you say) and seemed a bit of a cheat, but the practicalities made good sense.

The older property proved to be unsuitable. There were bats there, but it was simply too historic and not ordinary enough. It didn't fit the programme concept. The place was noisy and busy, busy, busy with jackdaws, and I thought that at least might be of interest; but nope, it just wasn't what they wanted.

Meeting Bill Oddie was easy and pleasant, but there was one thing that stuck at the back of my mind the whole time that I never quite felt like mentioning. I'd actually first encountered him in 1982, in the sense that I'd passed him on the staircase at the Royal Gala performance of *The Pirates of Penzance* at the Theatre Royal, Drury Lane. The evening was memorable, not least thanks to Tim Curry's gorgeously exuberant Pirate King dressed in purple breeches and knee-high leather boots, but it started when Steph, my zoo-keeper friend, and I met at the theatre door. She was clutching the crash helmet from her moped and I was pocketing my bicycle clips. The two of us had failed to take on board the 'Royal Gala' bit – you know, a red carpet and people in smart frocks. So we nipped across the road for a quick gin and tonic, aka Dutch courage.

We returned for the performance deep in conversation and headed up the stairs for the upper circle. Steph was talking very loudly (she always does) about a family scenario in Cumbria and she was watching her feet, but I'd seen two men standing chatting at the turning of the stairs, one of whom I recognized as Bill Oddie. And just as we passed, Steph proclaimed in full voice 'and my sister knows all about it, and he's an absolute ******** ********'.

There was no way the two gentlemen wouldn't have heard this unseemly description. Putting my hand on Steph's elbow, I propelled her up the stairs as quickly as possible. The embarrassment of that moment had stayed with me for years and I would have liked to have apologized. Somehow the opportunity to explain just never presented itself. Probably for the best.

CHAPTER 13

Teamwork

We'd been counting the bats as they emerged for nearly an hour, and we'd just passed the 1,000 mark. They were pouring out of the eaves of the cottage, either side of two gloriously colourful hanging baskets, and it was more than just a little hectic. Two teams of trainees were in place catching individuals in the long-handled nets and then carefully extracting them for inspection, while others were frantically clicking away on the tally counters. The stream of emerging bats was almost constant, and at times they seemed to tumble out of the eaves in their impatience to be off and away. As things began to slow down, the trainees with the tally counters were all muttering about aching thumbs. Is there such a thing as bat worker's thumb, they queried, and should future risk assessments for the site take this into consideration?

The total count was just under 1,300 soprano pipistrelles. When we had run the course the previous summer there had 'only' been around 600 bats, so a second colony had clearly arrived, more than doubling the numbers. That made it a spectacularly large and exciting roost by UK standards, and everyone involved was well and truly chuffed.

Truth be told, it was also another one of those smelly roosts. You don't get that many bats in a confined space without the pong, but fortunately this was a problem for the National Trust, not me. The cottage belonged to the Attingham Park estate in Shropshire, and all those attending were either National Trust wardens or other members of staff. The Trust owns a vast number of properties and hardly any of their buildings *don't* have bats of one sort or another, so training people in house for roost management was invaluable.

Between 1990 and 1993 the National Trust ran a total of 17 courses in conjunction with the VWT Bat Project, and by early 1994 some 340 Trust staff, including 180 from the building departments, had attended at least one course. Many of the wardens would go on to obtain licences.[1]

The mega-pip roost would have been a particularly memorable experience – it's certainly one of the largest I've ever counted – but I hope they remembered the hanging baskets too. The cottage was occupied by the head gardener from the estate and both baskets were beautiful, but one was even more spectacular than the other, the one that was closest to the bats' main access gap. In other words, it was the one that was getting a nightly shower of bat fertilizer. Here was the evidence that a bit of bat poo on your garden can work wonders.

There's another training course worth recounting, this time on the Isle of Skye. You have to be made of really stern stuff to be a Scottish bat worker. To begin with, there are the sunset times. Bats don't emerge from their roosts until after sunset. Some are more enthusiastic to get on the wing than others; the pips can pop out bang on sunset while others such as Natterer's and Daubenton's will take almost an hour or more before they feel inclined. In Scotland sunset in mid-June is almost 10.30 at night, hence many of the bats won't have even shown their heads before 11.15 or 11.30 pm. That makes for a very, very late night after counting the little blighters. We have it cushy in south Devon, where sunset is an hour earlier.

Secondly, there are the Scottish midges, which are ferocious. On the Isle of Skye we'd been standing outside a stone-built lodge waiting to see Daubenton's. As the evening darkened and the midges swarmed in horrific numbers, we were all forced to make an ignoble and hasty retreat to the minibus. It was not dissimilar to the Monty Python scene where everyone is shouting 'Run away! Run away!'

Teamwork comes in all forms. Yes, you see it on the best of the field trips and courses when everyone gels and mucks in together and when all the group members take turns lugging the ladder through the woods to get to the bat boxes. But for someone in my position, who's done relatively little over the years without the support and involvement of others, it comes in many other forms as well. It includes those people who have worked into the early hours of the morning to process the sound recordings from a bat detector that's been stuck in a barn all week, those who put their hands up to do a dawn survey at the very tail end of the season when the starting temperature registers a miserable four degrees, and others who've been at the end of a late-night telephone call for a second opinion on the latest crisis.

Sean is one of my regular surveyors. He's a laconic, willowy man in his early 30s with a dry sense of humour. He obviously thinks I'm completely barking but doesn't seem to mind too much, which frankly at my age is very uplifting. He's also very reliable, darn good with a bat detector and rarely moans unless I've forgotten to bring the wine gums and chocolates for the after-survey pow-wows (milk chocolates, mind you, not the dark stuff).

His patience has been tested, though. One evening we turned up to do an emergence survey of a redundant scout hut, a small rectangular timber structure that was due for demolition. I'd been there before so had a good idea of the bats we might find: just a few odds and sods that tucked themselves up behind the horizontal boarding at the back of the building.

'So this is where we've seen them,' I started to explain. 'They've all been on this side.' Pontificating further, I added, 'It's north facing and at this time of year they're almost certainly going to be males. Hey, look, here's a whiskered bat. Come see.'

Hah! This was a great opportunity for extra curriculum training, I thought. Sean had seen and recorded lots of species emerging and in flight but never seen one in the hand, not up close and personal. There at head height, several inches inside the edge of the board, was the bat with its ever so slightly scruffy dark brown fur with golden frosted tips, as if it had just had expensive highlights at the hairdresser's. I got my handy bat-extracting tool out, a dentist's mirror with a narrow extendable arm, and lengthening the arm I carefully nudged the bat downwards to where I could take it between my fingers.

'OK, here you go. Is it a male? Should be.'

He had it in his hand, and it was at that moment I realized I only had my long-distance glasses with me. I was stuffed without my other glasses.

'Er, it should have a penis,' I said, not being able to see anything in detail. 'Has it got a penis?'

He was concentrating and didn't reply.

'Look down there, is that a willy?'

'Get your finger out of the way, George.'

'Yes, but is it? I can't see a thing. It should have a willy. There, just there…'

'I know where the penis should be,' he snapped.

'Right. Yes, OK.' I shut up.

'It's a male. I can see that it's a male,' he concluded.

'Great, so what shape is it?'

'*What?*'

'It's important to check the shape. If it's sort of thin and pencil-shaped then that's typical of whiskered, but if it's bulbous that's more like a Brandt's bat…'

He stood staring at me.

'Probably best to put the bat back,' I suggested. 'It's not important right now.'

Teamwork spans time as well. In April 1994, wearing my voluntary roost visitor hat, I was asked to inspect the roof of a local residential care home where some chimney repairs were needed. It was a majestic, stone-built Victorian villa dating from the mid-nineteenth century with a tall, square roof with four ridge lines surrounding a central roof light. Sometime in the 1980s a smaller extension had been added to form a rear annex, and this had a separate roof void with its own ceiling hatch.

It was immediately clear that the care home was an important roost. The main roof was well used by grey long-eared bats and the annex was occupied by a big lesser horseshoe colony, both rare species. The live-in owners were delighted. Two months later I organized our first evening emergence survey, and we counted out 72 lessers and 18 of the grey long-eareds. The lessers emerged from multiple gaps in the dormer corners and the eaves in both east and west walls, and the long-eareds emerged from the eaves

Whiskered bat

of the tall south-facing wall, so we quickly learnt that we needed a minimum of four – ideally five – pairs of eyes to cover the activity.

In 1995 the BCT was running trials for their new National Bat Monitoring Programme (NBMP), so we again carried out summer surveys, this time following their protocols regarding dates and recording forms. Thirty years later, in June 2024, we were still counting the bats out as part of the NBMP. That's an awful lot of survey effort from many, many bat workers and even family and friends on some occasions. The lesser horseshoe bats have done well over the years and the colony now typically numbers around the 200 mark, although it did reach a peak of 242 one year. The typical number of greys present each summer was nearer 30 throughout the 1990s and up until 2006, but then it halved. We don't know why.

The NBMP processes the results for the all the colony counts, and these plus the results from other types of surveys using citizen science provide the data for population trends and patterns at the regional and national levels, which in turn can then inform conservation policy and management practice.

The greater horseshoe bats from the Brixham quarry caves have been counted and monitored by volunteers since the early 1990s (every summer barring 2007), and well over 40 people have done their bit to add to our knowledge of the site. Team members have come and gone, but there are stalwarts such as Chris Smallbones who now do it year after year.

It was Chris's father, Nigel Smallbones, who first introduced me to the caves in the autumn of 1990. He wandered up at a function at Paignton Zoo and simply said 'I've got bats.'

When some people say this, it sounds like a bit of an affliction. The way Nigel said it, it sounded like a grand boast, and he added, 'Do you want to come and see?'

He was the warden of the site and he knew he had greater horseshoe bats. He didn't know how many, what they were doing there or what it all signified in the great scheme of things, but he assumed that someone who worked on a bat project might be interested.

He was certainly right. South Devon was considered to be a stronghold for the species and Nigel's quarry sounded really interesting. So in the January and February of 1991 we did our first surveys with invaluable help from some local cavers, and later in the summer we sat outside the caves and did our first emergence count, all of which confirmed, respectively, the presence of good hibernating numbers and a breeding colony.

Two or three times each summer ever since, three or four of us walk down to the quarry gate, wearing our hard hats and head torches and carrying our detectors and tally counters. We tiptoe across the quarry floor, avoiding the rare lichen that carpets the ground, before climbing onto a pile of massive boulders, scaling up a couple of short, lumpy vertical sections and then scrabbling across the loose scree slopes towards the upper stone platform.

From there we have a choice of sitting outside the largest, most cavernous cave or perching at the entrance of the smaller one. Both have pros and cons. The larger cave is used by gangs of pigeons as well as the bats and the entrance stinks, but you can stand or sit without too much discomfort. The hole to the smaller cave is above a sheer sloping rock face, some scrubby buddleia and a narrow stone ledge. It's not as smelly outside, but it's too narrow to stand safely or without getting in the way of bats so you just have to crouch, parking one buttock on the ledge while twisting the rest of your body to have a view of the entrance.

The bats' behaviour isn't straightforward, however, and there's a heck of a lot of faffing about. There's a fair bit of 'light testing', as it's called, with bats suddenly appearing out of the smaller cave entrance and then shooting back down and in again, as if to say, 'oh no, not yet, not yet…' Or they emerge and dash off round the corner into the big cave, where they do several circuits before coming back to the smaller one. It's all in, out, in, out, shake it all about, and keeping track of numbers takes concentration.

When we think they've all finally shoved off and left the quarry we lower ourselves into the nursery cave and do a quick head count of any babies that have been born: tiny but big-faced and hideously ugly wriggling things that have yet to grow their wings to full length. Then we slowly and very carefully make our way down the rocks in the dark, trying desperately hard not to be the first one over the years who slips and breaks a bone.

Is it *enjoyable*, you might be asking yourself? It's definitely an experience, that's for sure, but it's not for everyone. There have been occasions when we've been in position in good time when the weather has been warm and still and the sounds of the gulls drift across the quarry; when there are insects flitting and buzzing on the buddleia near your face and it's a time for quiet contemplation and appreciation. Sometimes peregrines are present and their calls cut the air; more often there are tawny owls in the distance. And the bats themselves: you're seeing something dynamic, and you get to feel and hear the sound of air under their wings as they rush past. For most of us it's a thrill and a privilege.

And what has it told us? Well, the numbers themselves are mere numbers, but put them together with other information, for example from the radio-tracking project and from comparative studies,[2] and you get a bigger picture. Most of this indicates that the Brixham greater horseshoe roost is a pretty odd site. The limestone rock is like mouse cheese, so it makes for terrific hibernation conditions with lots of damp chambers and rifts, but it makes for a lousy breeding site – and if the females had ever bothered to read the textbooks they probably wouldn't be there.

The summer numbers have never shown the upward trend that most of the other sites in England and Wales have seen within the last three to four decades. When comparing some of the purpose-built and managed roosts to the caves, one can only image the bats' Trip Advisor comments:

> *Converted barn*: 'Great place to stay! Spacious, luxury accommodation and lots of modern fittings, including extra-grip linings under the roof, an anti-barn owl feature and even an electrically heated maternity box. Really appreciate the thoughtful and tasteful layout. The south annexe is a brilliant idea, perfect for the kids. Loved the location. Surrounded by countryside and loads of good eating places – check out the woodland rides and cattle pastures! Could stay here all year.'
>
> *Brixham cave*: 'Think there might have been something wrong with the heating. And the ceiling leaks after heavy rain. Great place to meet family and friends in the winter if you don't mind the trek over; it's a bit out of the way. Good sea views but don't bother with the town. Best food 4 kilometres away.'

And yet the bats return year after year, and while they do there will still be people to count them.

With big buildings or an extensive range of buildings, teamwork comes in the form of lots and lots of pairs of eyes and bat detectors. Members of the Somerset Bat Group came up trumps in the summer of 2009 for a memorable evening at Cleeve Abbey, a magnificent site near the pretty west Somerset village of Washford. The abbey comprises some truly beautiful structures, many remarkably intact. There's a fifteenth-century refectory with a vast carved timber angel roof, monastic cloisters and ground-floor chambers, a great dormitory and an imposing two-storey gatehouse. And just for good measure there's a later seventeenth-century farmhouse range that was added to the west side of the refectory.

The buildings are listed up to the eyebrows of the precious carved angels and are cheerfully occupied by bats galore. A minimum of seven species had already been recorded by the Somerset Bat Group from a visit in 2006. The refectory needed re-roofing, however, and this meant we had to get a better picture of which bats were where exactly before the works could take place. My initial visits, poking around here and there and with helpful tips from Ed and Helen Wells from the group, suggested that the bats were pretty much all over the place, and also that the farmhouse roof was occupied by a colony of our friendly and now familiar lesser horseshoe bats. Moreover, their droppings were almost everywhere throughout the abbey, including areas where there would soon be roof works and some ruddy great scaffolding.

Critically, we needed to know what the lesser horseshoes were playing at, how they were getting to and from the farmhouse roof, and what we could expect from the other odds and sods. We needed a cunning plan. If a bat so much as sneezed, I wanted someone to hear it or see it.

It seemed only logical to go to the bat group. Members had been there before, and any bat worker, voluntary or otherwise, would love this sort of gig. English Heritage approved a modest donation for the group's involvement and Ed and Helen organized the troops, 16 of them in total, with varying levels of experience, some with more than 20 years of bat work under their belts and others relative newcomers. The English Heritage staff mucked in as well, since we needed parking and access outside the normal opening hours. We even had permission for a couple of campervans to stay overnight, so a few of us could do some dawn observations and collect the sound recorders the following morning.

And where, exactly, did all these surveyors stand to implement this cunning plan? Well, as the organizer it's best not to put your assistants in the most precarious locations such as wobbly wall tops or next to great gaping holes in the foundations; that's what you're paid to do. You tend to develop an annoying habit of yelling 'Trip hazard!' a lot of the time, though. Fortunately, all of the positions at Cleeve were either at ground level or within the 'safe' parts of the building interior that were open to visitors.

After that consideration there's a bit of a knack to placing people in the right place, but there's also a great deal of luck in who sees what. A few unlucky members of the group that evening might have felt that they'd spent a long and less than stimulating time just staring at a lot of medieval masonry. The rest of us saw rather more action. That's always the way it goes. The best bit is always at the end when everyone puts their heads together with cups of coffee and chocolate biscuits and we plot the results as a whole. In this case that was a tally of at least 55 emerging bats of four different species and, most importantly, being at last able to map the route that the lesser horseshoe bats were taking.

It wasn't what you'd call a straight-line route; in fact it was downright convoluted. They first dropped down out of the side of the farmhouse roof into a lower, adjoining storeroom above the ticket office and shop; they flew through the gap at the east end of the store into the upper floor of the west range, through an arched doorway into the central hallway and up a spiral staircase to the second floor, where they followed the corridor through a doorway (which we later realized was a fire door that the staff kept permanently wedged open for the bats) and into the upper chamber. They flew to the top of the roof and over the wall of the chamber, then down again on the other side into the refectory, through a curving stone staircase and into the lodging chambers of the lower ground floor. There, they flitted about for a bit before deciding whether to exit the building via a wall slit in the north-east corner of the custodian's room or through several of the window slits in the south wall. The whole pattern of movement was absolutely extraordinary.

At dawn, although there were only a couple of us left to see what was happening, it was clear that the bats followed the same route in reverse back to the farmhouse roof.

The overall results allowed us to make valuable recommendations both in relation to the re-roofing works and for longer term management. The lesser horseshoes were given their own equivalent of a fire exit by modifying an existing roof light in the store, and the fire door into the upper chamber was fitted with a magnetic hold-open device, which meant it could be kept in an open position but would automatically close in the event of fire. The thoughtful but non-regulation wedge was removed.

The same cunning plan was used for Powderham Castle in the summer of 2019 with help from members of the Devon Bat Group. This is another huge, complicated and gorgeous building that is perpetually in need of upkeep and repairs. It stands in stately grounds to the west of the River Exe estuary, overlooking Lympston and Exminster to the east. From the mainline train route along the edge of the estuary you can see the fallow deer and the little egrets in the park, with the castle in the background.

In 2019 the estate needed to carry out a range of small-scale investigative works to parts of the upper roof fabric, including romantically named structures such as the Flagpole Tower, the Thirties and the Courtney Tower.[3] I'd gone poking around in the towers from the rooftop – not a unified, single rooftop but a multi-level, up-and-down and stepped, part-flat and part-pitched rooftop – and the signs of bat use were sparse and not what we might have considered to be well defined. It was almost certainly an incomplete picture, but the budget for the project was tight. So welcome, stage left, seven members of the Devon Bat Group and Colin, a licensed colleague, who could provide valuable and cost-effective eyes and ears for a large-scale survey.

It was a beautiful summer evening; clear, still and blue. Colin and I were to go on the rooftops. If anyone was going to fall off the roof it ought to me or Colin. Preferably not me. So the others were placed in the courtyards and in the gardens where they would have good and safe views.

The new Lord Devon, who had inherited the title at the death of his father in 2015, had volunteered to take the two of us through the castle to get onto the roof. We followed him up narrow stairs, through multiple doors and passages, large and small, and eventually into a tiny staff kitchen where he pointed to a small window. This wasn't quite what I was expecting. I'd taken a different route when I'd first been with the estate builders, and you really anticipate something a little bit more sophisticated in such a grand property. But no, it wasn't a joke, so I climbed up onto the sideboard by the sink and awkwardly scrambled outside. Why does so much of what we do seem so undignified? I thought. Colin was led off towards the north side of the castle.

Once on the roof I got my bearings and waved to several of the surveyors below, and Colin eventually radioed to say he was in position too. We were three storeys up. We both had good close views of the various bits of the towers and slate slopes that we needed to see, and the others would be covering everything below us. We waited.

The evening light had only just started to dim when I heard the first calls from noctule bats. They materialized high in the air over the trees and pond to the south of the castle, their loud 'chip-chop, chip-chop' calls coming from the bat detector. We often see noctules at the beginning of the evenings, one or two of them high above us, heading in a purposeful straight-line flight from their tree roosts towards the night's foraging area. But here they were, at least half a dozen flying and feeding at eye level straight in front of me, noisy as heck and spectacular to watch, swooping in sudden sharp dives and dramatic turns against the backdrop of the deer park and the setting sun.

Things were a lot quieter on the rooftop itself. Much, much quieter. In fact, by the time that the two of us had somehow, miraculously, managed to navigate our way back down and onto firm ground once again, and the team had reassembled, the total number of emerging bats recorded for the evening was two common pipistrelles. Technically that equates to 0.22 of a bat per surveyor, an utterly meaningless statistic – just to emphasize the point that it was a lot of surveyor effort for very little roosting activity.

But results are results, and just what was needed from a reporting point of view. Most of the surveyors did see the noctules, plus a few serotine bats on the wing and pipistrelles feeding in the rose garden. Someone also identified swifts nesting in the eaves of the castle walls, so it wasn't a bad way to spend a summer evening.

CHAPTER 14

I'll Just Have a Look in the Attic

People like me visit the homes of others to look for bats in the name of ecology and conservation. But it doesn't mean that we don't notice what else is going on, what the circumstances might be or what other sights there are to be seen. We've pretty much seen it all.

'Sorry about the mess.'

Why do so many people say this sort of thing when they open the door to their homes?

I've heard it over and over and over again across the years, and yet it doesn't necessarily seem to equate to the actual state of the place. Sometimes when it's said you can look over the person's shoulder into the interior and think 'Nah, this isn't a mess.' At other times you see the chaos and carnage unfolding behind them and think, 'Holy shit, they're not kidding.'

A safe and tactful reply has usually been 'Don't worry, I've seen worse.'

Which is eminently true.

The expressions of mess don't seem to cover cleanliness, which is a different matter, although they do go hand in hand. A home is easier to keep clean if every surface from carpet to window sill, to kitchen counter to shelving and sideboards isn't strewn with loads and loads of stuff. There's a broad threshold of tolerance to grubbiness. There are those who adore immaculately clean and spotless homes. Others, probably the majority, tolerate so much dust and then no more. Dust is only natural after all. And then there are those who cease to see the dirt around them – simply don't care or are no longer able to do anything about it. Perhaps like our cheerful and friendly cattle farmer and his filthy kitchen.

I've seen the full spectrum: the sparkling homes of the wealthy, the people who can afford a house-cleaning budget of £2,000 a month to keep the place looking expensive. And I've seen a wide bathroom window sill adorned with beachcombed shells and pebbles, fir cones and pot plants, covered in a heavy layer of dust, matted into a blanket-like coat by cobwebs and fungal threads. And a kitchen window sill similarly thick with years of greasy dust, yellowed by the aerosols of cooking oils. I fall into the middle bracket. Housework is boring and I'm content to live with some food crumbs on the carpet and a light layer of dust on the bedside table … but not much more.

No. It seems to me that the sentiments of mess refer to untidiness and clutter. Clutter, by definition, is a collection of things lying about in an untidy state, so the more stuff we have the greater the potential for clutter. And boy, do we have a lot of stuff.

Where does it all come from? The answer is complex and takes us into the realms of ancient anthropology and social history because, frankly, we've always had stuff and it's in our genes. We, *Homo sapiens*, turned up roughly 315,000 years ago, but we came from a long, long history of primate and early Hominid ancestors. Generally considered to be nomadic hunter-gatherers, our ancestors, so the experts believe, first acquired tools such as stone flakes and hammer stones some 2.6 to 3.3 million years ago.

I love the idea that way back then an early stone age man might have come back home to the missus one evening to the following scene:

> 'Hello darling, I'm home.'
> 'Hello sweetie, have you had a good day?'
> 'Yes, guess what? I've found some brilliant new stones to work on. I'll just put them at the back of the cave.'
> 'Oh, but honey, we already have a huge pile of stones.'
> 'Yes but you never know when they might come in handy.'

Why not? It probably started with stones and just grew from there. Solid stones and bones that fossilize last longer in the archaeological record than degradable elements, such as wood and fabric, but we can infer the acquisition of other tools from other evidence. Ancient cave paintings are variously dated at 25,000 to 30,000 years old, with the oldest in Indonesia at 45,500 years. Those artists needed the right implements to hold and apply the coloured dyes … more tools.

And what about clothes? A huge amount of information has come from Ötzi, the frozen man discovered in the Alps on the Austrian/Italian border in 1991. He's dated to be about 5,300 years old and DNA studies of his clothing showed him to have been wearing the leather hides from at least four different types of animal, including bear, roe deer and a domestic form of goat. And his shoelaces were made from cattle skin.

Older shoes have been found, mostly of leather but some with woven grasses, including a pair of sandals in a Spanish cave dating from around 6,200 years ago. I'd love to know when and where the first shoe rack appeared. My money is on the Egyptians, as they had a very natty line in sandals, but don't quote me on this.

But humans weren't just acquiring *useful* things. We started making things that we liked or that meant something important to us at a remarkably early stage. We had forms of adornment using materials such as shell 'beads', bones, feathers and coloured pebbles. We'd moved onto artistic creations by at least the early Stone Age, such as the iconic small stone female figurine called the Venus of Willendorf.

We're a socially complex species. We live long lives in diverse multi-generational families and societies. We demonstrate cultural exchange, loosely meaning the transference of learnt knowledge, beliefs and habits and possessions to others over time.[1] In wealthy societies, ancient or modern, we've learnt how to build ourselves sophisticated homes that we fill with our families and all the stuff that such families need – or don't need but just want – according to age, schooling, fashions, hobbies, social pressures and aspirations.

In theory, therefore, whatever we see in the living part of the house can overflow into the attic, but actually it doesn't work like that. The attic is another country; they do things differently there. For some, the attic becomes an extra playroom. Full flooring and raised tables are installed for a fantastic mini-world of miniature trains with stations and junctions and signals and all. I've only ever seen the empty tables after these owners have passed away or sold up. I've never seen the trains in all their glory, just the hollow roof space where they once rattled their way along the tracks. Photographic enthusiasts convert the attic or part of it to a darkroom. I've seen at least three that were used for growing marijuana, with foil-lined roof slopes and compost littering the floor. At each of these the new owners were understandably at great pains to explain that it was nothing to do with them and the police knew all about it.

You might think that the roofs of the largest and swankiest country houses are going to be a goldmine of storage items and belongings, but in fact they tend to be empty (except for bats) for several perfectly logical reasons. To begin with, most of them are of a certain age, which means they generally have high ceilings. Which means getting stuff in and out is too much of a faff unless someone has installed a staircase in the old servants' quarters or one of those eccentric ladder devices with ropes and pulleys.

I recall a particularly posh Georgian house near Dartmouth in the South Hams. Ceiling access to the main roof was via a top-floor bathroom, and *no one* was ever going to store anything in that roof. It was the biggest bathroom I'd ever seen with the highest bathroom ceiling I'd ever seen, and it required a two-person, full-length ladder operation to get to the hatch. Furthermore, the wallpaper upon which the ladder first had to be rested looked outrageously expensive, vivid with brilliantly coloured birds of all species, alive in posture and plumage. You did *not* want to mark that wallpaper. I could have luxuriated in the bath for hours admiring those birds. My colleague said they were too loud and they made him feel ill. There's no accounting for taste.

Anyway, if you live in a very, very large house you usually have enough spare rooms or outbuildings for all the boxes and paraphernalia that would normally go in the roof. My work has taken me to lots of large country houses, many listed, many held under the stewardship of the National Trust. Almost without fail the roofs support rich numbers of bats, but I can't think of one that was crammed with stored stuff or indeed housed anything other than old water tanks and piping.

In May 2006 I did find a single item in the roof of Greenway House, Agatha Christie's former home. The property was bequeathed to the Trust in 2000 and an extensive programme of structural repairs, restoration and caretaking was needed before it was opened to the public in 2009. The one item was a dusty, faintly mustard-coloured copy of *Woman's Weekly* from the 1950s. From memory I want to say 1956, but that may not be accurate. It had lain immediately inside the hatch entrance to the smaller of the two roofs, possibly untouched for 50 years. Every single item within the house was eventually catalogued so I hope it survived and that it might even be on display somewhere.

The 1940s newspapers in the old wartime bedroom with the jacket on the back of the door weren't unique, and I've found many over the years in roof spaces. They span the decades and are a time-warp back into history. And they give a useful insight into the length of time that a roof interior has been undisturbed. Occasionally, loose pages are just left lying around, too insignificant and innocuous to be noticed or removed. The oldest I ever came across were from December 1919 in a large, stone-built Victorian villa in Torquay. The solid walls had kept the rodents out of the roof and conditions were dry, so the pages just lay there for the best part of a century.

Newspapers were frequently used to line ceilings when loose insulation material was installed, such as vermiculite chippings. Astonishingly, as recently as June 2023 I came across a Dartmoor bungalow where the *only* insulation was straw, loosely and thinly scattered over the pages of the *Western Morning News* dated Monday 23 December 1946. Three things came to mind: a) by no stretch of the imagination did this meet any energy efficiency regulations; b) let's not even think about the fire risks; and c) the family must have been well 'ard to live there in the winter.

It's easy to think that if the interior of the house is tidy or chaotic the same will be encountered in the attic, but it doesn't follow. An utterly shambolic house can have a tidy and well-organized attic, and conversely a stylish, minimalist living space may hide cascading piles of junk above the ceiling.

The out of sight, out of mind mentality is very definitely a factor, but it's not just what we store; it's how we store it. There are two basic approaches.

First, you open the ceiling hatch and simply lob in the item that you don't want cluttering up the living space. It can seem the easiest solution, the most desperate or the laziest, or all of these. It's the favoured method in rental properties where the tenants have disappeared, leaving it all for someone else to sort out. Empty cardboard boxes; an unrolled sleeping bag and the pink knick-knacks that a small girl might have played with. Nothing of any worth any longer. Or you can see it with young couples who suddenly find themselves overwhelmed with babies and children and the avalanche of clothing and paraphernalia that this entails.

Secondly, you board out the roof space. You could start with a small section of boarding or demonstrate forethought and pre-planning by boarding out the entire roof. Some owners install shelving for multi-storey effect, while others rely on stacking – which can vary from the ergonomically efficient to downright haphazard.

In a modest 1990s house I visited in Dawlish a short while ago, the owners had managed to squeeze in 65 plastic storage boxes at one end of their roof. The boxes were in rows of five, six boxes deep. At the sides where the roof was lowest there was only enough height for two stacked boxes, but they were stacked three-high mid-slope and four-high under the apex. Sixty-five boxes in total. Respect – although I didn't see any signs of numbering or labelling. I hoped they'd know where to look should they wish to retrieve anything in particular…

This brings us to the actual contents. Where does it all come from? From longevity of family use certainly, for example in homes where the grandparents have lived for decades and where the attics contain not just their belongings but those of their children and even of their grandchildren. And then, conversely, in our own attics that we've inherited from our grandparents and parents.

There can be stacks of photo albums (photo-what? the young people ask) and loose photographs and framed photos that run through the generations. In the homes of couples with babies and young children, the hallway and kitchen walls shout with their proud, joyous photographs taken by the professionals, but the grandparents know they will soon get replaced with the school photos, the growing-up-photos and getting-married photos. All the old outgrown portraits move to the attic. My mother kept an entire trunk full of framed photographs of me and my four brothers from the 1960s and '70s, but photos from the decades that followed covered the walls and shelves.

There's crockery, china and crystal glass in some attics. I never ask about this sort of thing, but sometimes the information is offered and I'm told they're wedding gifts or 'it was in the family'. And flower vases; an awful lot of flower vases.

Valuable antiques? Almost certainly, though I'm not very good at spotting these. An old but very elegant rocking horse was difficult to miss in one roof.

There's all the stuff from the baby years. Potties top the list, but everything else can be seen: plastic baths, travel baskets and cots, pushchairs, prams, bouncers, baby carriers (with or without sunshades), playpens. There was once a lovely wooden high seat, clean and well cared for, which the grandparents had used for their children and grandchildren. They were keeping it in preparation for their first great-grandchild.

Outgrown toys and redundant board games litter roof spaces, too. A plastic Twister® mat always makes me smile. 'I used to play that as a kid!' I think, 'and then again with my own kids.' There are prize-winning rosettes, mostly horse-related but also one for 'Guinea Pig Grooming, Second Place'. Then there are the fruits of school and university years, sports prizes and trophies, and cardboard boxes labelled 'Jonathan. 2nd year Uni' and 'Sarah, London, 1983', and all their academic textbooks.

Heaps of magazines and collections of hobbyist publications used to be common, including those on trains, aeroplanes and vintage cars. And in the early years we'd come across an occasional porn mag too, but I'm glad to say I haven't found one of those for a while, presumably since the material is now so accessible online. Books are ubiquitous, clearly overflowing from the bookcases and shelves downstairs. One owner had installed shelving around the full circumference of the roof, turning it into their own full-blown library. It was an astonishing sight.

And then there are the attics that have become the depository for every conceivable form of sports equipment and outdoor gear. Climbing ropes, skis, wetsuits, paddles and surf boards. And yes, even a kayak once, just a small one. And leather jackets and crash helmets. So much of it from a life before having kids or getting old, except maybe the golf clubs, or from phases that the kids have since outgrown.

Mirrors and paintings of various descriptions get propped against the chimney. I often wonder where the pictures come from. Have the owners downsized so they no longer have enough wall space? Or is it a second (or third?) marriage where half of the new partnership has made it clear they dislike what came before them? So many people seem to think that an attic is an OK place to store clothes, which I find strange. The vacuum packed bags of baby and toddler items are maybe OK, but more often they simply get stuffed into loose plastic bags that fall open and carelessly spill their contents. There may be a crammed clothes rail that might be covered in a bit of protective sheeting, but usually not.

Every now and then there are perplexing sights and you just go 'Uhh?'. There was the roof stacked with a dozen jumbo packs of toilet rolls. Had the couple been panic buying during Covid? Did they manage rental properties or did they just have a thing about toilet rolls? We'll never know.

In another roof in early 2023 I found a cardboard box containing 12 bottles of tomato ketchup. That's a lot of ketchup. I asked, and there was a perfectly logical explanation. The owners had planned a big barbecue, which was cancelled because of Covid. Three years later I reminded them where they'd put the ketchup.

Do I need to mention the Christmas decorations and suitcases? Not really, but I have just for good measure.

We live in a materialistic world. We acquire many, many things during the course of our lives and a great number of these remain important to us throughout that time, but from what I've seen I believe we can be shockingly poor at taking care of our possessions. We often show little respect or regard for them; we store things casually, recklessly and lazily. A roof space can be a grotty place, damp and hostile or boiling in the summer. It may not have bats but it may well have rodents or spiders and flies that gnaw and poop and stain. So many things are damaged and spoiled when left unprotected and ignored for years on end.

We don't seem particularly good at passing on the things that we no longer want or find necessary; things that could very well be used and valued by others. We procrastinate until it's too late and then it's someone else's problem. Think of the woman who'd created her own mini-library in her attic. She passed away, leaving her poor daughter with the enormous task of emptying the roof. It's not at all uncommon.

As I think about retirement, I've wondered if I might make quite a good clutter consultant. Mine would be the brutal approach: 'Bin it. Bin it … Charity Shop. Bin it. Bin it. Charity Shop … Ooh, that's nice, can I have that?'

You might be wondering what's in my attic. Not bats, sad to say.[2] And not much else either is the honest answer. Divorced and relieved of many items that might have ended up as roof clutter, I bought a lovely little bungalow and eventually had the roof converted into a bigger, better bedroom, so the only space that remains is a small, low void at the back. But I can tell you exactly what it holds.

There are the sentimental memories from my childhood and early school years in the form of family albums and photographs of brothers and parents, all contained within a very old brown leather travelling case with the stamped initials 'H.M.', my grandfather, Henry Moreland. And there's a large shoebox full of postcards. The majority of these are from my father who sent them whenever he was away in the navy and working on fisheries duty with MAFF.[3] There are others from my brothers, husband, school friends and relatives. I cherish these, but they'll only go into the recycling box once I've passed away.

There's a clear plastic box full of children's books which I read to my sons. *Hubert's Hair Raising Adventure* was a favourite:

Hubert the lion was haughty and vain
And especially proud of his elegant mane.

And others with smile-worthy titles such as *Underwear!* and *The Lighthouse Keeper's Lunch*. I may have grandchildren one day.

There are books belonging to my oldest son from his late teens. It's definitely about time he reclaimed these but he's probably forgotten. His backpack's there too, but I've found it useful so he can't have that. There's one item of clothing: a wonderful elegant black evening dress by Blanes of London dating from the early 1960s and given to me by my godmother. I wore it for my 40th birthday party. It still fits and is in perfect condition. I kept it well wrapped. I really must take it to a vintage shop.

And the stuffed mole that was mentioned in the introduction? Yeah, that's mine. It's in another brown leather case with other stuffed mammal specimens. My husband gave me the briefcase when I first started on the Bat Project. I think he thought I was going to be an executive-type ecologist. It turned out I was more of an in-the-field and up-the-ladder sort, but it was perfect for keeping the animals that I used for teaching. I had them prepared by a taxidermist. There are a couple of rats, black and brown, a water vole and even an edible dormouse that someone sent me in the post. The *correct* way you send a dead animal in the post.

There are two other boxes with other mammal bits and pieces, including a small glass jar of yellow plastic beads, the sort used for feeding to badgers with sticky sweet peanuts; the plaster casts of otter and badger footprints; a couple of badger skulls, a fox skull and a roe deer antler; the baculum or penis bone of a stoat. Yes, it's a thing. And there's also some otter spraint and samples of deer droppings for identification.

Kumquat anyone?

Notes

Chapter 2: The Otter in the Wheelbarrow, and Welcome to the World of Bat Work

1. The wallaroo is a member of the kangaroo family. As its name suggests, it's midway in size between wallabies and the larger kangaroos.
2. Remember dieldrin; it'll appear again. The poisons affected many species including birds of prey such as the peregrine and sparrowhawk, which were poisoned through eating the seed-eating farm birds.
3. The NCC was later split into separate national bodies for England, Scotland and Wales with new names. There have been other name changes so what first became English Nature was later renamed Natural England, which is where we are today.
4. Mitchell-Jones and McLeish 2004; Mitchell-Jones 2004.
5. Mitchell-Jones, Hutson and Racey 1993.
6. Now Fauna and Flora International.
7. Macmillan 2015.

Chapter 3: Is a Bat a Rodent? Will it Eat Banana and Other Questions

1. For example, visit the BCT website.
2. Stebbings 1986.

Chapter 4: Meet the Owners

1. Peter was also employed by the VWT. One of his roles was to manage a large breeding colony of greater horseshoe bats at a roost in Buckfastleigh, near the River Dart in south Devon. We'll hear about these bats later.
2. Early eighteenth century if you need to ask. I know I did.

Chapter 7: The Joy of Ladders

1. Bat rabies, as it's sometimes called, differs from classical rabies, the sort we associate with dogs and other canines. This means that we still maintain a formal rabies-free status in the UK.

Chapter 8: Oh Crap!

1. Bryson 1990.
2. Wrap in soft paper, place in sealed plastic bags, then in a firm container. Post fastest delivery and make sure recipient is expecting it! It might be best to freeze before posting in some cases.

Chapter 9: Can I Help You, Officer?

1. See Robinson, Webber and Stebbings 2000.

Chapter 10: Batting from a Boat

1. The site was purchased by the VWT in 1988. At that time there were records of only 12 breeding roosts for greater horseshoe bats throughout England and Wales (Stebbings 1988).
2. The current status of the roost is unknown following reports of barn owl activity in 2024.
3. Robinson, Webber and Stebbings 2000.

Chapter 12: Call the Press!

1. Greenaway and Hutson 1990.

Chapter 13: Teamwork

1. Bullock 1995.
2. See Ransome 1997.
3. If you're into the history and architecture of ancient masonry, Powderham is really, really interesting, but sorry, I'm an ecologist, so you'll have to do your own background reading.

Chapter 14: I'll Just Have a Look in the Attic

1. Let's not get bogged down in the definitions of culture and cultural exchange: people argue about it. But see an interesting take on the subject in *The Cultural Lives of Whales and Dolphins* by Hal Whitehead and Luke Rendell (Whitehead and Rendell 2015).
2. I've never seen any signs of bats in any of the homes where I've lived in the last few decades. Feels like a bit of an insult really.
3. Ministry of Agriculture, Fisheries and Food.

References

Bryson, B. (1990). *The Mother Tongue: English and How It Got That Way*. Penguin Books.
Bullock, D.J. (1995). 'Bats in the National Trust: a preliminary review'. *Biological Journal of the Linnean Society* 56: 119–26.
Greenaway, F., and Hutson, A.M. (1990). *A Field Guide to British Bats*. Bruce Coleman Books.
Macmillan, H. (2015). *From Mallards to Martens*. The Vincent Wildlife Trust.
Mitchell-Jones, A.J. (2004). *Bat Mitigation Guidelines*. English Nature.
Mitchell-Jones, A.J., and McLeish, A.P. (2004). *Bat Workers Manual* (3rd edition). JNCC, Peterborough.
Mitchell-Jones, A.J., Hutson, A.M. and Racey, A.P. (1993). 'The growth and development of bat conservation in Britain'. *Mammal Review* 23: 139–48.
Ransome, R.D. (1997). 'The management of greater horseshoe bat feeding areas to enhance population levels'. English Nature RR no. 241.
Reason, P.F. and Wray, S. (2023). 'UK Bat Mitigation Guidelines: a guide to impact assessment, mitigation and compensation for developments affecting bats. Version 11.' Chartered Institute of Ecology and Ecological Management. Ampfield.
Robinson, M.F., Webber, M., and Stebbings, R.E. (2000). 'Dispersal and foraging behaviour of greater horseshoe bats, Brixham, Devon'. English Nature RR no. 344.
Stebbings, R.E. (1986). *Which Bat Is It?* The Mammal Society.
Stebbings, R.E. (1988). *Conservation of European Bats*. Christopher Helm.
Whitehead, H. and Rendell, L. (2015). *The Cultural Lives of Whales and Dolphins*. The University of Chicago Press.

Index

African spot-necked otter, 8
Alcathoe bat, 2, 19
aldrin, 8
Angus Bat Group, 13
asbestos, dangers of, 47
Avon Bat Group, 12

Barbastelle, 2
Bat Conservation Trust (BCT), 13
bat detectors, 27
bat(s)
    DNA techniques, 19–20
    eating habit questions, 17
    question about, 16–23
    reproduction, 4
batting from a boat, 74–77
    River Dart, 75–76
    River Mardle, 75–76
Bechstein's bat, 50
Berry Head, 71
Bill Oddie's How to Watch Wildlife
    series, 87
Brixham caves, 76, 93
brown long-eared bat, 25, 53, 70
Bryson, Bill, 56
Burton, John, 11

carpets and nappies, 35–43
    disgusting atmosphere, 38–41
    dog hairs, 42
    footwear use, 36–37
Chanin, Paul, 38, 40
Chapman, Peter, 24, 38, 74
Chiroptera order, 16
Christie, Agatha, 105
clutter, 102
Clwyd sites, 12
common pipistrelle, 3, 19, 25, 66, 100

'dangly' bats, 2
Dartmoor National Park Authority, 50

Daubenton's bat, 54–55, 85
Devon Bat Group, 84, 99
Devon sites, 19
dieldrin, 8, 66
DNA techniques, 19–20
    testing of bat droppings, 57, 60
dog issues, 46–47
droppings, *See* poo
dung pit, 57

European Lyssavirus disease, 54

Fauna and Flora Preservation Society
    (FFPS), 11
Field Guide to British Bats, A
    (Greenaway and Hutson), 84
fixed ladder, issues with, 49
Flagpole Tower, 99

Goldsmith, John, 12
greater horseshoe bat, 85–86, 93
    from the Brixham caves, 76
    in Devon, 85
Greenaway, Frank, 84
grey long-eared bat, 41–42, 60

Habitats Regulations, 13
hibernation sites of bats, 3
    Norfolk, 12
home owners, 24–34
    anxiety problems of, 26
    calls from, 25–26
    distress of, 25
    hostile owners, 26, 30
    restoration, repair works, 31
    smell problem, 28–29
    voluntary roost visitor and,
        relationship between, 29–30
    warmness of, 28–29, 32–33
home visits, things to watch, 101–110
    attics, 103–104, 107

Index

books, 110
crockery, 107
mirrors and paintings, 108
newspapers, 105
photo albums, 107
potties, 107
social well-being, 103
untidiness and clutter, 102
wealth, 102, 104
horseshoe bats, 2, 58, 79
Hubert's Hair Raising Adventure, 110
human evolution, 102–103
clothes, 103
footwear, 103
tools, 102
Hutson, Tony, 11, 83

Izzard, Eddie, 86

joy of ladders, 46–55

Kestrel, 22
Kuhl's pipistrelle, 20

ladders and roof access, 48–49
Lane, Rupert, 51
Leisler's bat, 12, 20, 84
Leptospirosis (Weil's disease), 65
lesser horseshoes, 99
Lighthouse Keeper's Lunch, The, 110
lindane, 66
London Zoo, 7
'Moonlight World' at, 7
long-eared bats, 58, 70
Lord Devon of Powderham Castle, 44–45

Markham, Steve, 76
media coverage on bats, 82–88
Bill Oddie's How to Watch Wildlife series, 87
radio interviews, 85
TV channels, 85–86
Mitchell-Jones, Tony, 9
Morris, Pat, 9
Mother Tongue, 56
Myotis family, 2, 19

Nathusius' pipistrelle, 19–20, 83
National Bat Monitoring Programme (NBMP), 93
Natural Resources Wales, 14
Nature Conservancy Council (NCC), 8
night-time bat work, 78–81
noctule, 1, 21, 100

Oddie, Bill, 87–88
organochlorine insecticides, 8
otter in the wheelbarrow, 7–15
Otter Trust in Suffolk, 8
outdoor ladders, 50

Paignton Zoo in Devon, 13, 83, 95
parvovirus, 65
pentachlorophenol (PCP), 66
pipistrelles, 2, 3, 5, 6, 19, 20–22, 25, 27, 53, 66, 75
Pirates of Penzance, The, 88
poo, 56–67
  DNA analysis, 57, 60
  dung pit, 57
  health risk from, 65
  horseshoe bats, 58
  identification of, 58
  'latrine', 57
  long-eared bats, 58
  reveals behaviour and ecology, 57
  spraints, 56
Powderham Castle, 99
press attention to bats, 82–88. *See also* Media coverage on bats

Reason, P.F., 10
River Dart, 75–76, 86
Robinson, Mark, 71, 72–73, 76
rodents, poo, 57–58

Savi's pipistrelle, 20–21
Schrieber's bent-winged bat, 10
serotine bat, 53–55, 100
Smallbones, Chris, 94
Smallbones, Nigel, 94
Somerset Bat Group, 12, 96–97
soprano pipistrelle, 18–19, 28, 54, 66, 89
spraints, 56

Taylor, Rob, 51
teamwork, 89–100
   spans time, 92
   VWT Bat Project, 90
toxocariasis (roundworm), 65

Underwear!, 110

Vincent Wildlife Trust (VWT), 11
   VWT Bat Project, 13, 24, 90

voluntary bat workers, 25
voluntary visits, 48

Wayre, Philip, 8
Webber, Maurice, 76
Weir, Vincent, 14
whiskered bat, 51, 91–93
Wildlife and Countryside Act 1981
   (WCA 1981), 8, 11
Wray, S., 10